The Cosmic Breath

Somewhere in the Darkness of the Cosmos, Volume 2

SANDEEP CHAVAN

Published by SANDEEP CHAVAN, 2025.

THE COSMIC BREATH

First edition. February 7, 2025.

ISBN: 979-8230769408

Written by SANDEEP CHAVAN.

Table of Contents

Table of Contents

To the Vedic Scholars and Physicists—

The seekers of truth who have spent lifetimes unraveling the mysteries of the cosmos, whether through the sacred verses of ancient wisdom or the precision of scientific equations.

May this book serve as a bridge between **the wisdom of the past and the discoveries of the future**, inspiring new dialogues in our eternal quest to understand the universe.

"That which is far beyond is also within. That which dissolves is also reborn. The cosmos breathes, and in its breath lies the rhythm of creation and dissolution."— **Inspired by the Rigveda and Modern Cosmology**

"Look deep into nature, and then you will understand everything better."— **Albert Einstein**

Preface

The universe has always been a mystery—a vast, ever-changing expanse that holds the secrets of our existence. From the earliest human civilizations to modern astrophysics, we have sought to understand its nature, its origins, and its ultimate fate. This book, *The Cosmic Breath: Black Holes, Pralaya, and the Energy Cycles of the Universe*, is an attempt to bridge the ancient and the modern, weaving together the wisdom of Vedic cosmology with the discoveries of contemporary physics.

For centuries, science and spirituality have been seen as opposing forces, yet the deeper we explore both realms, the more we realize that they speak of the same truths in different languages. The ancient concept of *Pralaya*—the cyclical dissolution and rebirth of the universe—bears striking similarities to modern theories of cosmic evolution, black hole physics, and cyclic universe models. This book is an exploration of these profound connections.

As an engineer, researcher, and educator, I have always been fascinated by the intersections between disciplines—how insights from one field can illuminate mysteries in another. The parallels between Hindu cosmology and modern astrophysics are too compelling to ignore, and through this work, I seek to present these ideas in a way that resonates with both the scientifically inclined and those drawn to the philosophical depth of ancient traditions.

Throughout this journey, we will explore the structured energy models of the cosmos, the role of black holes as regulators rather than mere destructive entities, and the possibility that our universe is engaged in an eternal cycle of expansion and contraction—what I call the *cosmic breath*. In doing so, we will challenge outdated notions of singularities, examine the implications of dark energy as a structured force, and reconsider the nature of time itself.

This book is not just about astrophysics or metaphysics—it is about **redefining our perception of reality.** It is a call to embrace a more holistic view of the universe, one that does not separate science from wisdom, but rather integrates them into a unified framework of understanding. Whether you are a scientist, a philosopher, or simply someone curious about the nature of existence, I invite you to join me on this exploration of cosmic cycles, hidden forces, and the eternal dance of creation and dissolution.

In the vast, silent void of space, the universe breathes. Let us listen to its rhythm.

Sandeep Chavan

1: The Universe Unfolded – A Journey Across Time and Energy

The universe has always been an enigma, an infinite expanse that stretches beyond human perception, yet deeply influences every aspect of existence. From the flickering stars in the night sky to the rhythmic waves of time that shape civilizations, everything in the cosmos follows a pattern—one of creation, preservation, and dissolution. But does the universe itself abide by this rhythm? Is it born from a singular event, destined to expand endlessly, or does it breathe in cycles, collapsing into itself only to be reborn?

These questions are not new. Ancient Indian sages, through deep contemplation, envisioned the universe as a self-sustaining, cyclical entity, with time flowing like an ever-turning wheel. Modern physics, armed with equations and telescopic eyes, now inches closer to similar conclusions. Theories of the Big Bang, black holes, and cosmic entropy hint at a universe that may not simply expand forever but could eventually contract, dissolve, and emerge anew—mirroring the concept of Pralaya described in the scriptures.

In this journey, we explore how these two seemingly different approaches—the intuitive wisdom of the ancients and the empirical rigor of modern science—lead us to the same cosmic truths. Are black holes the agents of cosmic destruction, or do they play a deeper role in maintaining the balance of existence? Is Pralaya just myth, or could it be a forgotten scientific insight encoded in symbolism? As we unfold the mysteries of the universe, we step into a realm where energy, time, and cosmic cycles intertwine, offering a profound perspective on existence itself.

1.1 The Eternal Question: Does the Universe Have a Beginning or an End?

For as long as human consciousness has pondered existence, one question has persisted: Is the universe eternal, or does it have a definitive beginning and end? Does everything we see around us emerge from a singular point of creation, only to expand forever into an uncertain future? Or does the cosmos follow a grand cycle, where everything that exists will one day dissolve, only to be reborn again?

Ancient Indian philosophy and modern science have both wrestled with this question, albeit from different vantage points. The Vedas and Puranas speak of an infinite cosmic rhythm—Brahma's day and night, where creation and dissolution are but two phases of an eternal cycle. In contrast, for much of its history, Western science leaned towards a linear perspective, assuming a universe that had a fixed beginning, possibly in the Big Bang, and an ultimate fate of endless expansion or final collapse.

Yet, as scientific understanding deepens, the notion of a cyclic universe is emerging once again. The idea that the cosmos breathes—expanding and contracting in a never-ending loop—mirrors the Vedic concepts of Yugas, Kalpas, and Pralaya. While physics speaks of entropy and gravitational collapse, Indian cosmology describes the dissolution of the elements into the unmanifest (Avyaktam), only for them to reorganize and reform into a new reality.

Despite their differences, both approaches—one rooted in ancient wisdom and intuition, the other in mathematics and observation—converge on one profound truth: Nothing truly ends. The universe, like existence itself, is an ongoing transformation. Whether through scientific theories of the Big Crunch, Big Bounce, or black hole information conservation, or through spiritual descriptions of Pralaya and cosmic rebirth, the conclusion remains the same—the universe is not a one-way journey, but a cycle of energy, balance, and renewal.

1.2 Time and Energy: The Twin Forces That Shape Reality

In the vast expanse of the cosmos, two forces silently govern everything—Time (Kala) and Energy (Prana). These are the fundamental threads weaving the fabric of reality, shaping the birth and dissolution of stars, galaxies, and even human existence. Ancient Indian thought recognized Kala and Prana as eternal principles, dictating the rhythm of the universe, while modern physics, through relativity and quantum mechanics, has begun to unravel their intricate relationship.

Kala (Time) and Prana (Energy) in Vedic Thought

Vedic cosmology describes Kala as relentless, absolute, and unidirectional—it flows unceasingly forward, bringing change and transformation. Prana, on the other hand, is the vital force, the infinite energy that pervades the cosmos. It is the force that animates all existence, from the smallest atom to the vast celestial bodies. Together, they form the two unchanging constants of the universe—Time dictates movement, and Energy fuels existence.

The Bhagavad Gita echoes this cosmic balance when Krishna declares:

"कालोऽस्मि लोकक्षयकृत्प्रवृद्धो"

(I am Time, the great destroyer of worlds).

This suggests that Kala is the ultimate force of transformation, dissolving everything in its wake, yet ensuring the continuity of existence.

Modern Physics and the Nature of Time and Energy

Science now tells us that time is not an absolute, fixed entity but a relative dimension, influenced by gravity and motion. Albert Einstein's Theory of Relativity revealed that time can slow down or speed up depending on gravitational fields. Near a massive cosmic object, such as a black hole, time can stretch infinitely, creating the illusion of a timeless zone—an idea that eerily resonates with ancient descriptions of dissolution (*Pralaya*).

Energy, in modern physics, is recognized as the underlying essence of reality. The famous equation $E=mc^2$ states that mass and energy are interchangeable, implying that everything, from the stars to human consciousness, is simply a form of energy in constant motion. This aligns with the Vedic concept of Prana, which states that energy is never destroyed but only transformed, much like the modern law of energy conservation.

Can Black Holes Distort and Regulate Both Time and Energy?

Black holes, the most enigmatic objects in the cosmos, seem to embody the dual control of Kala and Prana.

- Time slows down near a black hole due to intense gravity, similar to how Kala is said to dictate the lifespan of the universe.
- Energy is trapped, compressed, and possibly recycled—mirroring the ancient belief that Pralaya dissolves all matter into an unmanifest state, only for it to re-emerge in a new cosmic cycle.
- The event horizon acts as a boundary beyond which time ceases to have meaning, much like how Pralaya marks the transition between existence and non-existence.

This raises an intriguing question: Are black holes cosmic regulators of Time and Energy, just as the Vedas describe Kala and Prana as the forces governing universal cycles? If modern physics suggests that black holes could be key players in cosmic rebirth, ancient Indian thought already spoke of this balance through Pralaya and cosmic regeneration. The interplay between Kala and Prana, Time and Energy, Black Holes and Cosmic Cycles suggests that the universe is not just a collection of stars and matter but a carefully orchestrated dance of forces, regulating existence beyond human comprehension.

1.3 The Cosmic Cycle: Creation, Dissolution, and Rebirth

The universe is not a static entity. It evolves, transforms, and, perhaps, even resets itself in an eternal loop. While modern science is still unraveling the mysteries of cosmic beginnings and endings, ancient Indian wisdom has long envisioned the universe as part of a grand cosmic cycle—a rhythmic flow of creation (Srishti), preservation (Sthiti), and dissolution (Pralaya).

This idea is deeply embedded in the Hindu concepts of Yugas and Kalpas, which describe the passage of time in vast cosmic scales, akin to a universal clock ticking across ages and eons. As we delve into these cycles, we begin to see parallels with modern astrophysical theories, especially the Big Bang, the Big Crunch, and the role of black holes in cosmic rebirth.

The Cosmic Clock: Yugas, Kalpas, and the Cycles of Time

In Hindu cosmology, time does not move in a straight line; rather, it moves cyclically, like a great wheel (Kala Chakra). According to the Puranas, the universe undergoes repeated cycles of manifestation and dissolution, each spanning unimaginable time scales.

- A Kalpa (4.32 billion years) is considered a single day in the life of Brahma, the creator.
- At the end of each Kalpa, dissolution (Pralaya) occurs, resetting the universe into an unmanifest state before the next cycle begins.
- Four Yugas (Satya, Treta, Dwapara, and Kali Yuga) define the decline of cosmic order within each cycle, leading to destruction and renewal.

This concept aligns closely with modern discussions in cosmology, where scientists propose that the universe may not expand indefinitely but could one day contract and collapse into itself, leading to a new cycle of creation.

Scientific Perspectives: The Big Bang, The Big Crunch, and Beyond

The modern Big Bang Theory suggests that the universe began from a singular point of infinite density, expanding outward to form everything we see today. However, what happens next remains a matter of speculation.

- Some scientists propose that if gravity overcomes expansion, the universe will contract in a "Big Crunch," collapsing back into a singularity—a process eerily similar to Pralaya.
- Others propose a cyclic universe model, where the death of one universe seeds the birth of another. This aligns with the idea that Pralaya is not an end but a transition into the next cosmic phase.

Even more intriguing is the idea that black holes may play a crucial role in this cosmic rebirth.

Are Black Holes Gateways to a New Phase of Cosmic Rebirth?

Black holes, the most mysterious entities in the universe, appear to act as cosmic recyclers, absorbing matter, bending time, and possibly transporting energy across cosmic dimensions. Could they be the mechanism behind cosmic regeneration?

- Ancient texts describe Pralaya as a dissolution where all matter merges into an unmanifest state. A black hole, too, seems to pull everything into an unknown state beyond its event horizon.

- Some theories suggest that the matter consumed by a black hole could emerge elsewhere—possibly in a new universe. This mirrors the Hiranyagarbha (Golden Womb) concept, which describes the reformation of the cosmos after dissolution.
- Just as Pralaya ensures that no energy is truly lost, modern physics speculates that black holes might not destroy information but transform it in ways we do not yet understand.

Could it be that black holes are not just cosmic voids but gateways between one universe and the next? If so, then ancient seers and modern physicists may be describing the same truth—only in different languages.

The Cosmic Dance of Destruction and Rebirth

The Hindu trinity—Brahma (Creator), Vishnu (Preserver), and Shiva (Destroyer)—symbolizes this eternal cycle of the universe. Shiva's Tandava, the great dance of cosmic dissolution, represents the destruction necessary for renewal—a concept mirrored in black holes, which consume matter only for new stars and galaxies to form elsewhere.

Whether through myth or mathematics, through intuition or equations, the message is clear:

- The universe does not move toward a permanent end.
- It transforms, resets, and renews itself, driven by the forces of time and energy.

This raises a profound possibility—black holes may not be the graveyards of the universe but the very seeds of its rebirth. If Pralaya is the great reset, then black holes may be the cosmic conduits through which existence is rewritten, over and over again.

1.4 Ancient Indian Sages and the Search for Cosmic Truths

Long before telescopes scanned the skies and equations attempted to decode the universe, ancient Indian sages sat in deep meditation, gazing inward to understand the cosmos. They saw the universe not as a static entity, but as a living, evolving phenomenon—a dynamic interplay of energy, space, and time. Unlike the materialistic approach of modern science, which relies on observation and measurement, these sages sought truth through intuition, contemplation, and metaphysical inquiry.

Their vision of the cosmos was far ahead of its time. The universe, according to them, was not created once and for all—it was a grand cycle of emergence, transformation, and dissolution. From their profound insights emerged texts that described the mystical origins of the universe, the forces that govern its existence, and the eventual collapse into nothingness before rebirth. These ideas, once considered mythological, now find startling echoes in modern astrophysical theories.

The Nasadiya Sukta: The Hymn of Creation and the Birth of the Universe

One of the most enigmatic and profound texts of ancient Indian thought is the Nasadiya Sukta, found in the Rig Veda. Often referred to as the Hymn of Creation, it is a poetic yet deeply philosophical attempt to describe the origins of the cosmos.

It begins with a paradox:

"There was neither non-existence nor existence then.
There was neither the realm of space nor the sky beyond.
What covered it? Where was it? In whose keeping?
Was there then cosmic waters, in depths unfathomed?"

This passage eerily resembles the scientific description of the pre-Big Bang state, where neither time, space, nor matter existed—only a formless, undefined potential. The text continues:

"Then there was neither death nor immortality,
Nor was there the torch of night and day.
That One breathed, windless, by its own impulse.
Other than that, there was nothing beyond."

This "One" could be compared to the singularity before the Big Bang, where all matter and energy were compressed into an undefined, infinite point. The hymn also raises an astonishingly modern question—did even the gods know how creation happened? Or is it beyond all knowledge? This resonates with the contemporary debate in physics—can we ever truly understand the universe's origin, or is it fundamentally unknowable?

Through its mystical yet analytical approach, the Nasadiya Sukta presents a vision of the universe as an ever-changing field of energy and possibility, rather than a fixed creation with a beginning and an end.

Did Ancient Seers Intuitively Understand Cosmic Balance, Energy Equilibrium, and Gravitational Phenomena?

Many ancient Indian concepts align remarkably well with modern physics, suggesting that the sages had an intuitive grasp of the fundamental principles of cosmic balance.

1. Prana as the Universal Energy Field

- The concept of Prana (universal life force) is strikingly similar to modern theories of dark energy, which permeates the cosmos and governs its expansion.
- Just as physics describes energy as never destroyed but only transformed, Vedic teachings suggest that Prana flows through all existence, reshaping and renewing it continuously.

2. The Fabric of Space-Time and Akasha (Ether)

- Ancient Indian texts speak of Akasha, the formless, all-encompassing substratum of existence.
- This resembles the modern idea of space-time as an invisible fabric that bends, stretches, and governs the movement of celestial bodies.

3. Gravitational Equilibrium and the Cosmic Dance of Balance

- The Tandava of Shiva, often described as a dance of destruction and creation, could symbolize the gravitational forces that maintain equilibrium between cosmic expansion and collapse.
- Indian cosmology speaks of the universe "breathing"—expanding and contracting in cycles—which is now a concept in theoretical physics as the cyclic universe model.

4. Black Holes and the Concept of Pralaya

- Ancient texts describe Pralaya as a phase where all matter dissolves into the unmanifest (Avyaktam), only to be reshaped into a new creation.
- This aligns with the idea that black holes consume matter, possibly leading to the birth of new universes elsewhere.

Bridging Ancient Wisdom and Modern Science

Were these ancient insights mere poetic imaginations, or did they hint at an understanding far beyond their time? Without modern instruments or equations, how did sages conceptualize a self-regulating, cyclical universe that aligns with cutting-edge astrophysics today?

Perhaps the answer lies in the nature of knowledge itself. Science observes and quantifies, while ancient wisdom perceived the universe as an experience—one that could be understood through deep contemplation and alignment with cosmic rhythms. The sages may not have had telescopes, but their intuitive grasp of balance, energy, and transformation remains astonishingly relevant in the modern quest for understanding the universe.

As we move forward, we must ask—should science look back at ancient knowledge not as mythology, but as a different way of perceiving cosmic truths?

1.5 The Black Hole Paradox: Cosmic Destruction or Cosmic Balance?

For centuries, black holes were imagined as cosmic abysses—regions where gravity is so strong that nothing, not even light, can escape. Modern physics, rooted in Einstein's general relativity, describes them as gravitational traps, where space-time bends infinitely inward, leading to a singularity. To most, they are the ultimate void, the graveyards of stars, the end of matter and time.

But what if we have misunderstood their true nature?

Ancient Indian cosmology does not describe destruction as an absolute end, but as a transition—an energy redistribution that resets the balance of existence. In this sense, black holes may not be cosmic tombs, but cosmic regulators, ensuring energy equilibrium within the universe. Could it be that what we call "destruction" is merely an act of transformation on a universal scale—one that aligns with the Vedic concept of Pralaya?

Beyond the Abyss: Are Black Holes Energy Regulators?

Modern science suggests that black holes do more than just consume matter.

- They release high-energy radiation in the form of Hawking

radiation, subtly influencing their surroundings.

- They seem to play a role in galaxy formation, with supermassive black holes found at the centers of nearly every galaxy, possibly orchestrating cosmic structure rather than simply erasing it.
- Some theories propose that black holes may be recycling matter and energy into new cosmic formations, similar to the dissolution and rebirth cycle of Pralaya.

Instead of seeing black holes as the end of physics, we might need to view them as key players in the ongoing transformation of the cosmos.

The Event Horizon and the Vedic Concept of Dissolution (Pralaya)

The event horizon of a black hole is the boundary beyond which nothing can return—an eerie parallel to Pralaya, the phase of total dissolution in Vedic cosmology. In Hindu texts, Pralaya is not destruction in the way we fear it; it is the absorption of everything back into an unmanifest state, where time ceases to function, and matter dissolves into its fundamental essence.

Consider these similarities:

- The event horizon marks the boundary between existence and the unknown, much like Pralaya's role as the threshold between one cosmic cycle and the next.
- Time inside a black hole slows down infinitely—mirroring how Pralaya suspends all motion and progression before the next creation.
- Matter inside a black hole is compressed into a singularity, just as Pralaya describes all existence merging into Brahman before a new creation cycle begins.

What if black holes do not erase matter but instead return it to a cosmic seed state, similar to the Hiranyagarbha (Golden Womb), which births new universes?

Resetting Universal Energy Dynamics: The Role of Black Holes in Cosmic Cycles

If black holes are not the ultimate destroyers but cosmic reset mechanisms, then they may be the real architects of balance in the universe.

- Pralaya resets cosmic energy by dissolving matter into an unmanifest state. Could black holes be doing the same—absorbing excess cosmic energy and feeding the next phase of creation?
- Some theories suggest that black holes could lead to white holes in another universe, meaning that matter falling into a black hole here may emerge as new creation elsewhere.
- Could black holes be the cosmic lungs of the universe, breathing in excess matter and energy, regulating equilibrium, and ensuring that the cycle of birth and rebirth never truly stops?

If so, then black holes are not cosmic accidents, but cosmic necessities. They might be the Pralaya of modern physics—the balancing force that ensures the universe does not spiral into chaos, but instead follows an eternal cycle of transformation.

The Cosmic Balancing Act: Pralaya, Black Holes, and the Nature of Reality

The paradox of black holes—where matter is destroyed yet never truly lost—mirrors the paradox of Pralaya, where the universe dissolves but never ceases to exist. Perhaps, in looking at black holes, we are glimpsing the very mechanism by which the cosmos renews itself.

The question is no longer whether black holes represent destruction—it is whether we are ready to see them as a bridge between ancient wisdom and modern science, revealing the eternal cycle of existence.

1.6 Science and Spirituality: Two Paths to the Same Truth?

For centuries, science and spirituality have been viewed as opposing forces—one rooted in observation, logic, and empirical validation, while the other is based on intuition, experience, and transcendental wisdom. Science is seen as rigid and rational, while spirituality is often regarded as mystical and subjective. But when it comes to the cosmic mysteries of existence, time, and energy, are they really different? Or are they merely two languages attempting to describe the same truth?

Both modern physics and ancient Indian cosmology grapple with the same fundamental questions:

- Where did the universe come from?
- What is its true nature?
- Does it have an end, or does it cycle eternally?
- What is the role of energy, space, and time in shaping existence?

The answers, whether derived from mathematical equations or spiritual contemplation, seem to converge toward a unified understanding of a cyclical, interconnected, and ever-evolving universe.

The Scientific Method vs. Intuitive Spiritual Inquiry

Science relies on the physical senses, observation, and measurement to uncover truths about the universe. It seeks to quantify reality through mathematics, experiments, and testable theories.

Spirituality, particularly the kind practiced in ancient Indian traditions, takes a different approach—it depends on deep contemplation, inner realization, and intuitive wisdom. Instead of relying on external instruments, it turns inward to explore the nature of existence and consciousness itself.

Yet, despite their differences, both approaches recognize the same fundamental principles:

- The universe is governed by laws and patterns (science calls these natural laws, while spirituality refers to cosmic rhythms and Dharma).
- Reality is interconnected and dynamic, where everything influences everything else.
- Time, space, and energy are not absolute but are relative, fluid, and shaped by larger forces.
- The known universe is just a fraction of a much deeper, hidden reality (science explores this through dark matter, quantum fields, and black holes, while spirituality speaks of higher dimensions, Akasha, and the unseen fabric of existence).

The more physics advances, the more it starts resembling spiritual insights. Quantum mechanics, for example, suggests that particles exist in multiple states simultaneously until observed—a concept eerily similar to the Vedic understanding of Maya (illusion), where reality is shaped by perception.

The Limits of Mathematics in Understanding Infinite Cosmic Cycles

Mathematics has been the language of science, helping us define gravitational forces, atomic structures, and even the bending of space-time. But can mathematics truly describe the infinite, cyclical nature of the cosmos?

Some of the greatest challenges in modern physics stem from the incompleteness of mathematical models:

- The singularity problem in black holes suggests an infinite density where known laws break down.
- The multiverse hypothesis speculates on multiple realities but remains unproven.
- Theories like Loop Quantum Gravity and String Theory attempt to unify quantum mechanics and relativity, yet still lack definitive proof.

Ancient spiritual traditions, on the other hand, do not depend on numerical equations but rather on holistic, conceptual models to explain existence. The cyclical nature of Kalpas, Yugas, and Pralaya does not require a precise equation—it is an experiential truth, observed over time through deep meditative insight.

This does not mean that mathematics is flawed; rather, it may be incomplete when it comes to describing dimensions beyond our observable universe. If we truly wish to understand cosmic cycles, we may need to go beyond numbers and incorporate a more holistic perspective—one that includes intuition, consciousness, and cosmic rhythms.

Merging Ancient Cosmology with Modern Astrophysics

If we remove the boundaries between scientific knowledge and spiritual wisdom, a profound realization emerges:

- The Big Bang and Pralaya may describe the same event—a transition between one cosmic phase and the next.
- The laws of physics and the concept of Dharma (cosmic order) both suggest that the universe follows governed principles rather than randomness.
- Black holes, once considered cosmic anomalies, may instead be structural elements of energy balance, echoing the Vedic idea of Shiva's Tandava—the dance of destruction and creation.

So, what happens when we combine these perspectives? We get a richer, deeper understanding of the universe—one that is not just physical but also metaphysical.

By uniting science and spirituality, we open new possibilities:

- Understanding consciousness as a fundamental part of the universe, not just a biological accident.
- Exploring how energy transformation (whether through physics or spiritual ascension) dictates the flow of time and existence.
- Recognizing that the universe is not just expanding, but evolving through cycles—mirroring the same principles found in Pralaya.

Bridging Two Worlds: A New Path for Cosmic Understanding

For too long, science and spirituality have been treated as separate disciplines, but perhaps they were never meant to be apart. The greatest minds in history—from Nikola Tesla to Albert Einstein, from Adi Shankaracharya to Swami Vivekananda—have recognized that truth is neither purely scientific nor purely spiritual, but an integration of both. If we wish to truly understand the cycles of the universe, the role of energy, time, and cosmic balance, then we must embrace both the precision of science and the depth of spiritual wisdom.

Because in the end, whether through a telescope or meditation, through equations or intuition, we are all searching for the same thing: the fundamental nature of reality, and our place within it.

2: The Hidden Nodes of Cosmic Energy – Rahu, Ketu, and Black Holes

The universe has always been filled with hidden forces—some visible, others only sensed through their effects. In the previous chapter, we explored how black holes, once considered cosmic voids, may actually be energy regulators, balancing the flow of matter and time across the cosmos. But long before modern physics began decoding these mysteries, ancient Indian cosmology described two invisible entities—Rahu and Ketu—believed to govern celestial energy and disrupt cosmic order.

Rahu and Ketu, often referred to as shadow planets, have been depicted as cosmic forces responsible for eclipses, distortions in celestial motion, and shifts in human consciousness. Unlike the tangible planets that revolve around the Sun, these two have no physical form—yet their influence is undeniable. Similarly, black holes do not emit light or possess a defined surface, yet their immense gravitational pull shapes galaxies, bends space-time, and regulates cosmic energy flow.

Is it possible that Rahu and Ketu were early conceptualizations of space-time anomalies, akin to black holes and gravitational nodes? Could the ancient seers, through observation and intuition, have detected invisible cosmic forces that science is only now beginning to understand? In this chapter, we delve into the striking parallels between these shadowy celestial entities and black holes, seeking to bridge the wisdom of the past with the discoveries of the present.

2.1 The Eclipses of the Cosmos: Shadow Planets and Event Horizons

Throughout history, eclipses have been regarded with both awe and mystery. In ancient India, the disappearance of the Sun and Moon during an eclipse was not seen merely as a celestial event but as the work of two invisible cosmic entities—Rahu and Ketu. Unlike the physical planets that move in predictable orbits, these two were described as shadow planets—forces that disrupted the harmony of the cosmos, creating temporary darkness before allowing light to return.

In modern astrophysics, we now recognize another celestial entity that exhibits a similar behavior—black holes. These enigmatic objects, like Rahu and Ketu, have no visible form, yet they exert immense influence over the cosmic order, consuming light, bending space-time, and shaping the very structure of galaxies. Could these ancient descriptions of shadowy cosmic beings that devour light have been an early insight into the nature of event horizons and gravitational singularities?

The Mythological Significance of Rahu and Ketu in Vedic Thought

The story of Rahu and Ketu comes from the great Hindu epic, the Mahabharata, particularly in the episode of the Samudra Manthan (Churning of the Ocean). According to legend:

- The Devas (celestial gods) and the Asuras (titans) were churning the cosmic ocean in search of Amrita, the nectar of immortality.
- When the nectar was finally obtained, the gods devised a plan to keep it from the Asuras.
- One of the Asuras, Swarbhanu, disguised himself as a god and managed to sip the nectar.
- However, Vishnu decapitated him with his Sudarshan Chakra before the nectar could pass beyond his throat.
- Since he had already tasted the nectar, he did not die, but his head and body separated.
- The head became Rahu, and the body became Ketu—two

celestial beings that vowed vengeance on the Sun and Moon, who had exposed his deception.

- As a result, Rahu and Ketu continue their pursuit of the Sun and Moon, occasionally swallowing them, causing eclipses, before they escape again.

While this is often interpreted as mythology, it remarkably mirrors an observable astrophysical phenomenon: the periodic disappearance of celestial bodies into an unseen force, much like objects falling into the gravitational pull of a black hole.

Rahu and Ketu as Light Consumers: A Parallel with Black Holes

Just as Rahu and Ketu are said to "consume" the Sun and Moon, black holes do something eerily similar—they trap light within their gravitational grasp, making it vanish from the observable universe.

- A black hole's event horizon is the boundary beyond which light cannot escape, creating the illusion that it has been "swallowed."
- In an eclipse, the Sun or Moon appears to disappear temporarily—an effect that, in ancient times, was explained as Rahu or Ketu consuming them.
- Both Rahu and Ketu, as well as black holes, are "shadowy" forces—unseen but incredibly powerful, influencing the motion of celestial bodies.

In modern terms, the lunar nodes—the points where the Moon's orbit intersects the Earth's orbital plane—are responsible for solar and lunar eclipses. These nodes shift over time, creating predictable patterns of eclipses, much like the predictable nature of black hole accretion and gravitational lensing.

Could it be that Rahu and Ketu symbolized an early understanding of these invisible cosmic forces, functioning as unseen regulators of celestial mechanics?

The Event Horizon of a Black Hole and the Shadow Effect of Rahu and Ketu

A black hole's event horizon is a boundary beyond which light ceases to exist in the observable universe. This creates a shadow effect, making the black hole itself invisible except for the distortions it causes in nearby space. This behavior bears striking similarities to the role of Rahu and Ketu in eclipses:

- Just as the event horizon marks the region where light is absorbed but not emitted, Rahu and Ketu represent points in the sky where the Sun and Moon vanish temporarily.
- The EHT (Event Horizon Telescope) image of a black hole revealed a dark circular shadow surrounded by glowing matter—similar to the description of Rahu and Ketu as entities that obscure celestial bodies but do not permanently erase them.
- In Hindu cosmology, Rahu and Ketu exist in the shadows, never visible themselves but causing profound effects on celestial motion—just as black holes, though unseen, dictate the movement of stars and even entire galaxies.

This parallel suggests that the ancient seers might have intuited the presence of cosmic energy distortions long before modern astrophysics formally described event horizons, gravitational lensing, and singularities.

Conclusion: Were Rahu and Ketu Early Metaphors for Black Holes?

While the story of Rahu and Ketu comes from mythology, its deeper meaning aligns with scientific discoveries about light-consuming celestial phenomena. Whether interpreted as mythological figures or as symbolic descriptions of astronomical realities, these shadowy entities seem to hold a deeper truth:

- Both Rahu-Ketu and black holes act as unseen cosmic regulators, influencing celestial order.
- Both consume light temporarily, creating shadow effects in the cosmos.
- Both exist in a liminal space—neither fully part of the visible universe, nor completely separate from it.

As we move forward, we will explore whether Rahu and Ketu, like black holes, serve a greater role in cosmic energy flow, potentially acting as universal balance points, vortexes of energy regulation that influence the very fabric of space and time.

2.2 The Ancient Perspective: Unseen Vortexes of Energy Flow

In the previous section, we explored how Rahu and Ketu function as cosmic shadow forces, influencing celestial events much like black holes. But the deeper question remains: Why did ancient Indian seers describe invisible, formless entities with immense power? Were they merely mythological constructs, or could they represent an intuitive understanding of unseen cosmic energy flows, akin to modern space-time distortions?

While the physical planets of our solar system are observable and predictable, Rahu and Ketu exist only as abstract points, yet their impact is undeniable. This mirrors the behavior of gravitational nodes, energy vortexes, and black holes, which remain unseen yet shape the structure of the universe.

Why Ancient Texts Describe Rahu and Ketu as Invisible Yet Powerful Forces

Unlike other celestial bodies, Rahu and Ketu do not have a physical presence—they are defined only by their effect on other objects. In Vedic cosmology, they are depicted as:

- Unseen yet powerful, capable of swallowing the Sun and Moon.
- Forces that disrupt equilibrium, causing energy shifts and turbulence in the cosmos.
- Regulators of celestial order, affecting planetary motion and human consciousness.

These descriptions bear a striking resemblance to modern astrophysical concepts:

- Black holes are not directly visible but bend light and time, distorting everything around them.
- Gravitational waves, predicted by Einstein and detected in recent years, travel unseen through space-time yet influence massive celestial bodies.
- Dark matter, which comprises most of the universe, cannot be seen but dictates the motion of galaxies.

Could it be that ancient seers perceived energy distortions in space and described them in the form of Rahu and Ketu—celestial beings that exist beyond ordinary perception yet shape cosmic order?

The Vedic Understanding of Energy Imbalance and Cosmic Turbulence

In Vedic cosmology, the universe is seen as a dynamic field of energy interactions. Nothing remains static—forces constantly shift, realign, and create periods of harmony (Sattva) and turbulence (Tamas).

- Rahu represents chaos and illusion (Maya)—its presence disrupts balance, much like how black holes warp time-space

and absorb energy.
- Ketu symbolizes detachment and transformation—it dissolves old structures, akin to how black holes consume matter and convert it into pure energy.

This is similar to how gravitational distortions near black holes create turbulence in surrounding cosmic structures. When a star nears the event horizon of a black hole, its energy and mass are violently redistributed—a cosmic process resembling the disruptive and transformative roles of Rahu and Ketu.

Vedic texts also describe the movement of Rahu and Ketu as unpredictable and chaotic, influencing planetary alignments and cosmic cycles. This description mirrors:

- The way gravitational waves disrupt the space-time fabric, influencing cosmic motion.
- The behavior of black hole mergers, which generate enormous energy fluctuations and ripple through the universe.
- The unpredictability of quasars and cosmic jets, where energy flows chaotically before stabilizing.

These parallels suggest that the sages were not merely imagining celestial chaos—they may have been describing energy imbalances and vortexes within the cosmic framework, much like modern physics does today.

Can These Descriptions Be Interpreted as Early Observations of Space-Time Distortions?

One of the most fascinating aspects of modern physics is the realization that space and time are not fixed absolutes but dynamic, flexible constructs shaped by gravity and energy. The effects of black holes, gravitational lensing, and cosmic energy flows reveal that the very fabric of reality can be bent, twisted, or even punctured by unseen forces.

Ancient Indian texts suggest a similar understanding:

- Akasha (the ether-like cosmic fabric) was seen as a fluid field of energy where celestial motion occurs.
- Rahu and Ketu were described as "shadows" that moved through space, creating distortions in light and planetary influence.
- Pralaya (cosmic dissolution) was not a random event but the result of energy reaching a critical imbalance, much like black hole singularities leading to extreme gravitational collapse.

Modern interpretations suggest that black holes create "holes" in the space-time continuum, where energy behaves unpredictably. In Vedic thought, Rahu and Ketu distort cosmic rhythms, much like energy vortexes or gravitational wells.

If we strip away the metaphorical language and look at the underlying principle of unseen cosmic regulators influencing energy flow, the connections become even stronger:

- Just as Rahu and Ketu were believed to affect time cycles (Kaalachakra), black holes affect time dilation, slowing down or accelerating the passage of time in their vicinity.
- Rahu was said to create illusions, leading beings into Maya, just as gravitational distortions around black holes create mirages of light (gravitational lensing), making distant galaxies appear warped or duplicated.
- Ketu was associated with detachment and liberation, much like how black holes consume matter, stripping it of its previous form and transforming it into pure energy before ejecting it back into the universe.

Were these merely poetic descriptions, or did the ancient sages perceive energy structures and cosmic distortions that align with our modern understanding of astrophysics?

Conclusion: Were Rahu and Ketu Early Descriptions of Space-Time Distortions?

The deeper we analyze Vedic cosmology, the more it seems that Rahu and Ketu may have been early metaphors for cosmic energy nodes, gravitational influences, and unseen vortexes in space-time.

- Their invisible yet powerful influence mirrors the effects of black holes, dark matter, and gravitational waves.
- Their ability to distort time and light aligns with modern astrophysical discoveries of gravitational lensing and space-time warping.
- Their role as cosmic regulators in energy balance suggests that ancient thought recognized the presence of unseen cosmic forces shaping celestial motion.

Science may now provide mathematical equations to describe what the ancients perceived intuitively—but in the end, both approaches seem to be describing the same universal truth: that the cosmos is a dynamic, self-regulating system governed by hidden forces.

As we move forward, we will explore whether black holes, much like Rahu and Ketu, act as universal balance points, controlling the movement of energy and maintaining the equilibrium of existence.

2.3 Black Holes as Energy Regulators in the Cosmic Balance

Black holes are often viewed as cosmic anomalies—regions of extreme gravity that trap everything, even light, within their event horizons. Yet, as our understanding of the universe deepens, they appear less like chaotic voids and more like fundamental regulators of cosmic balance.

Much like Rahu and Ketu, which are invisible yet exert tremendous influence over celestial bodies, black holes govern the flow of energy, time, and gravitational forces across the universe. Their existence challenges our perception of reality, as they reshape the very fabric of space-time and dictate the movement of galaxies, much as Rahu and Ketu dictate the shifting energies of cosmic cycles. Could it be that black holes serve as natural energy regulators, maintaining equilibrium in the universe, much like these ancient shadow entities?

How Black Holes Manage Energy, Time, and Gravitational Forces

Unlike stars, which radiate energy outward, black holes accumulate and redistribute energy in ways that defy conventional astrophysical expectations. Their immense gravitational pull is not merely destructive—it reshapes cosmic structures, regulates the flow of matter, and controls the evolution of galaxies.

1. Regulating Galactic Motion:

- Black holes are found at the centers of nearly all galaxies, including our own Milky Way.
- Their presence ensures that stars and cosmic debris move in stable orbits rather than dispersing into space.
- Much like Rahu and Ketu, which govern planetary alignments and cosmic rhythms, black holes create gravitational anchors, preventing the universe from descending into disorder.

2. Time Dilation and Energy Transformation:

- As objects fall toward a black hole, their experience of time slows relative to the rest of the universe—a phenomenon predicted by Einstein's general relativity.
- This mirrors the idea that Rahu and Ketu distort the perception of reality, affecting the flow of time in Vedic

astrology.

- The extreme conditions near a black hole allow energy to be converted into different forms, similar to the concept of Pralaya, where energy dissolves and is restructured into new cosmic patterns.

3. Recycling Cosmic Matter:

- Black holes do not simply consume matter—they also emit powerful jets of high-energy particles, redistributing energy across galaxies.
- This resembles the role of Ketu, which signifies detachment and transformation—absorbing, restructuring, and releasing energy into new forms.
- In this sense, black holes are not just cosmic devourers; they act as catalysts for creation, much like Rahu and Ketu regulate celestial balance through eclipses.

The Concept of Gravitational Equilibrium—Similar to Rahu and Ketu's Role in Balancing Cosmic Cycles

One of the most fundamental laws of the cosmos is equilibrium—the delicate balance between forces that govern the motion of celestial bodies. Ancient Indian cosmology recognized this balance in the form of Rahu and Ketu, which, despite being disruptive, ultimately maintain the cosmic cycle of light and shadow, stability and transformation.

This is strikingly similar to the way black holes maintain gravitational equilibrium in the universe.

- If gravity were too strong, matter would collapse into itself, leading to an unstable universe.
- If gravity were too weak, galaxies would not hold together, and stars would drift apart into the cosmic void.

- Black holes act as stabilizing forces, concentrating energy in a way that ensures the structured evolution of galaxies and cosmic matter.

In Vedic thought, Rahu and Ketu exist at two opposing points, balancing each other's influence. This mirrors the concept of gravitational equilibrium, where black holes and cosmic expansion counteract each other to maintain order in the universe.

Could this be why ancient seers depicted Rahu and Ketu as existing in exact opposition, creating a stable yet dynamic interplay of forces? If so, black holes may serve a similar function—preventing the universe from falling into chaos by acting as cosmic regulators of mass and energy.

Could Black Holes Be Maintaining Energy Flow Across Cosmic Structures, Just as Rahu and Ketu Redirect Celestial Movements?

Black holes do not exist in isolation—they influence everything around them, shaping cosmic evolution in ways that extend beyond mere gravitational attraction. This raises an important question: Are black holes actively maintaining energy flow across the cosmos, similar to how Rahu and Ketu redirect celestial energies?

1. Energy Redistribution:

- The immense jets of energy emitted by black holes transport matter across vast distances, influencing the formation of stars and planets.
- This is reminiscent of Rahu and Ketu shifting the cosmic energies through eclipses and celestial alignments, impacting both the physical and metaphysical worlds.

2. Interplay Between Black Holes and Cosmic Expansion:

- The expansion of the universe is counterbalanced by the pull

of gravity from supermassive black holes at galactic centers.
- This mirrors the balance between Rahu and Ketu, which do not allow unchecked expansion or contraction but instead regulate movement and transformation.

3. Connection to Dark Matter and Dark Energy:

- Some theories suggest that black holes may be interacting with dark matter, an invisible substance that governs most of the universe's mass.
- In Vedic thought, Rahu and Ketu represent unseen but powerful influences, much like dark matter remains invisible yet dictates the structure of the cosmos.

If black holes are indeed regulating cosmic balance, then their role extends far beyond destruction. Much like Rahu and Ketu shape the cycles of light and shadow, black holes may be ensuring that cosmic energy continues to flow, allowing the universe to maintain its structure over billions of years.

Conclusion: Are Black Holes the Rahu and Ketu of Modern Science?

While modern physics and ancient cosmology use different languages, they appear to describe the same fundamental truth—that unseen forces regulate the cosmic balance, ensuring the universe does not descend into chaos.

- Rahu and Ketu manipulate energy through celestial alignments, black holes manipulate energy through gravity and space-time distortions.
- Rahu and Ketu are shadow forces that cannot be seen, yet shape cosmic order—black holes are invisible, yet structure galaxies and regulate energy flow.

- Both serve as cosmic regulators, ensuring that equilibrium is maintained in the grand design of the universe.

Perhaps, when the ancients spoke of Rahu and Ketu, they were not just describing eclipses and celestial omens—they were pointing toward a deeper reality, one that modern physics is only beginning to understand.

In the next section, we will explore how gravitational lensing, energy distortions, and the warping of light near black holes mirror the ancient descriptions of Rahu and Ketu as forces of illusion and hidden knowledge.

2.4 The Mathematical and Mystical: Bridging Ancient and Modern Knowledge

For centuries, humanity has tried to decode the universe through both mathematical reasoning and spiritual intuition. While modern physics relies on equations, ancient civilizations, particularly Vedic thinkers, relied on symbolic representations of cosmic forces. Surprisingly, the deeper we explore the phenomena of black holes, energy distortions, and gravitational lensing, the more we find striking parallels between scientific observations and ancient mystical descriptions.

In particular, Rahu and Ketu—the shadow planets that influence celestial mechanics without a physical form—bear a remarkable resemblance to black holes and gravitational distortions. Could it be that ancient astronomers observed the effects of space-time warping, gravitational energy shifts, and unseen cosmic forces, but articulated them in symbolic and mythological terms?

How Gravitational Lensing Around Black Holes Mimics Rahu and Ketu's Shadow Effect

One of the most fascinating aspects of modern astrophysics is gravitational lensing—a phenomenon where the intense gravity of a black hole bends light around it, distorting and magnifying the view of objects behind it. This effect makes black holes appear as shadowy voids surrounded by rings of bent light, much like the visual impact of an eclipse caused by Rahu and Ketu.

Comparing the Two Concepts:

Aspect	Gravitational Lensing (Black Holes)	Rahu & Ketu's Shadow Effect
Light Bending	Gravitational pull bends and distorts the path of light, creating the illusion of multiple or warped images.	During an eclipse, the Sun or Moon appears "swallowed" and then re-emerges, mimicking the temporary bending or disappearance of light.
Event Horizon & Cosmic Shadows	The black hole's event horizon is a region of space where light ceases to escape, creating a dark void in the sky.	Rahu and Ketu are mythologically described as "swallowing" celestial bodies, making them temporarily invisible.
Unseen Yet Influential	Black holes are invisible but shape cosmic structures by their gravitational force.	Rahu and Ketu have no physical form but influence planetary alignments and eclipses.

Just as gravitational lensing allows us to see objects hidden behind a black hole, ancient texts describe Rahu and Ketu as forces that momentarily obscure celestial bodies, only for them to reappear. This suggests that early astronomers may have been observing real distortions in space-time, interpreting them through the lens of cosmic balance and celestial energy.

Did Ancient Astronomers Detect Energy Distortions but Interpret Them in Symbolic Terms?

Modern astrophysics provides powerful tools—telescopes, mathematical models, and gravitational wave detectors—to measure the invisible forces at play in the universe. But before these technologies existed, ancient civilizations had only the naked eye, deep observation, and intellectual speculation to understand celestial mechanics.
Yet, their insights were surprisingly accurate in many ways:

1. Recognizing Unseen Celestial Influences:

- Ancient Indian astronomers knew about Rahu and Ketu's effects on eclipses long before modern science explained the concept of lunar nodes—the points where the Moon's orbit crosses the Earth's ecliptic plane.
- This suggests an understanding of gravitational interactions and celestial alignments, albeit expressed in symbolic language.

2. Describing Warped Space-Time in Metaphors:

- The Vedas often describe the cosmos as a constantly shifting, illusionary field of energy—a notion that mirrors modern discussions about the warping of space-time near massive cosmic objects.
- The Bhagavad Gita and Upanishads speak of a reality that is shaped by perception and hidden forces, similar to how gravitational lensing alters our view of the universe.

3. Understanding the Cyclic Nature of Cosmic Forces:

- Ancient texts suggest that Rahu and Ketu do not permanently consume celestial bodies; they only obscure them for a brief period before releasing them.
- This mirrors the idea that black holes do not destroy energy but transform and redistribute it elsewhere.

Could it be that ancient scholars detected subtle cosmic distortions but lacked the mathematical framework to describe them scientifically? Instead, they might have personified these forces, weaving them into narratives that aligned with their broader philosophical and metaphysical understanding of the universe.

If Rahu and Ketu Represent Shifting Nodes of Influence, Do Black Holes Serve a Similar Role in Reshaping Cosmic Energy?

In Vedic cosmology, Rahu and Ketu are considered points of energy concentration, existing not as physical bodies but as shifting vortexes of cosmic influence. They are responsible for energy fluctuations, celestial transformations, and disruptions in linear motion.

Similarly, black holes serve as energy nodes in the universe, controlling the movement of galaxies, absorbing excess mass, and regulating cosmic structures.

- Black holes act as attractors, pulling in nearby matter and shaping the energetic flow of the universe.
- Their immense gravity bends time and space, much like Rahu and Ketu alter the course of planetary alignments.
- They redistribute energy in ways we do not yet fully understand—just as Rahu and Ketu are said to guide the hidden forces of celestial motion.

Both black holes and Rahu-Ketu function as "hidden hands" of the cosmos, influencing the grand design without being directly observable. In this sense, black holes may be the modern scientific counterpart to these ancient cosmic regulators.

Conclusion: The Convergence of Science and Ancient Thought

When we examine ancient texts alongside modern astrophysical theories, we find stunning similarities between the symbolic descriptions of Rahu-Ketu and the properties of black holes and gravitational distortions.

- Rahu and Ketu obscure celestial light, much like black holes bend and distort light through gravitational lensing.
- Both concepts represent unseen forces that shape the cosmos without direct visibility.
- Ancient astronomers may have observed energy distortions and explained them in a way that resonated with their spiritual and metaphysical worldview.

As we continue to decode the mysteries of black holes, cosmic equilibrium, and the unseen architecture of the universe, we may find that ancient wisdom had already touched upon truths that science is only now beginning to quantify.

In the next section, we will explore how Rahu, Ketu, and black holes serve as veils between realities, hinting at the deeper, hidden nature of the universe itself.

2.5 The Cosmic Veil: Rahu, Ketu, and the Nature of Reality

Reality, as we perceive it, is shaped by what we see, what we measure, and what we believe exists. Yet, the universe is filled with unseen forces—energies that influence existence but remain beyond direct observation. In Vedic cosmology, Rahu and Ketu are not just celestial entities affecting eclipses; they are symbolic of illusion (Maya), hidden truths, and the cosmic veil that separates the known from the unknown.

Modern astrophysics, too, deals with such unseen mysteries. Black holes, with their event horizons and gravitational warping, hide information, bend reality, and challenge our fundamental understanding of space and time. The deeper we explore the universe, the clearer it becomes that Rahu, Ketu, and black holes may all represent different facets of the same cosmic principle—the veil that obscures ultimate reality.

Rahu and Ketu: Guardians of Maya and the Hidden Forces of the Cosmos

In Vedic thought, Maya (illusion) is a force that distorts reality, preventing beings from perceiving the ultimate truth of existence. Rahu and Ketu are deeply connected to this concept because:

- They cause eclipses, momentarily concealing celestial bodies from view.
- They create illusionary effects, much like gravitational distortions that bend light and space-time.
- They influence perception and transformation, guiding individuals through periods of uncertainty, just as black holes govern cycles of cosmic energy redistribution.

Rahu, associated with desire, ambition, and illusion, symbolizes the pursuit of material existence, where one gets trapped in worldly distractions, much like an object falling into a black hole, unable to escape.

Ketu, representing detachment and transcendence, is the force that dissolves illusion, allowing one to move beyond the veil of Maya into higher awareness—similar to the idea that information is never truly lost in a black hole but undergoes transformation.

This duality of Rahu and Ketu, existing as opposites yet complementary forces, mirrors the paradoxical nature of black holes, which are both destructive and creative, concealing yet revealing hidden truths about the universe.

The Parallel with Black Holes: Cosmic Veils and the Concealment of Information

Black holes act as cosmic veils, concealing information and preventing direct observation. According to Einstein's General Relativity, anything that crosses the event horizon of a black hole is lost to the outside universe—similar to how Rahu and Ketu create temporary cosmic blindness during eclipses.

Key Parallels between Black Holes and Rahu-Ketu:

Concept	Rahu & Ketu (Vedic Cosmology)	Black Holes (Modern Physics)
Illusion & Hidden Reality	Rahu symbolizes material illusion, while Ketu represents spiritual detachment from illusion.	Black holes conceal information behind their event horizon, making it impossible to see what lies within.
Disrupting Perception	Eclipses caused by Rahu-Ketu obscure celestial bodies, temporarily distorting cosmic order.	Black holes distort space-time and bend light, making objects appear warped or invisible.
The Boundary of the Unknown	Rahu-Ketu represents points of transition, where energy shifts between material and spiritual realms.	The event horizon is a boundary beyond which information and matter behave in ways that defy conventional physics.
Transformation and Loss of Identity	Ketu dissolves ego and attachment, leading to transcendence and transformation.	Black holes transform matter and energy into unknown states, possibly transporting it elsewhere.

The most fascinating connection between these ancient and modern concepts lies in the paradox of information loss and transformation. In Vedic philosophy, nothing is truly lost—only veiled until realization occurs. Similarly, modern physicists, including Stephen Hawking and Leonard Susskind, have debated whether information that enters a black hole is lost forever or remains encoded in some unknown form, waiting to be retrieved.

Could this mean that black holes are not just cosmic voids but gateways to deeper realities, much like Rahu and Ketu symbolize the thresholds between illusion and enlightenment?

Are Black Holes the Ultimate Veil, Hiding the True Nature of Space, Time, and Matter?

The study of black holes has led scientists to profound realizations about the hidden nature of the universe:

1. The Breakdown of Space-Time:

- As objects approach a black hole, time slows down due to extreme gravity.
- Beyond the event horizon, time as we understand it ceases to function, raising questions about the true nature of existence beyond observable reality.

2. The Mystery of Singularities:

- At the core of a black hole lies a singularity, where known laws of physics collapse.
- This is similar to how Rahu and Ketu disrupt normal planetary motion, causing unpredictable energy fluctuations.

3. The Multiverse and Cosmic Gateways:

- Some theories suggest that black holes could be gateways to parallel universes, transporting matter and energy into new dimensions.
- This aligns with the Vedic idea of cosmic cycles, where dissolution (Pralaya) leads to the rebirth of new realms beyond what is currently perceivable.

If black holes hide the fundamental truths of the cosmos, just as Rahu and Ketu obscure celestial bodies, could it mean that the universe itself is layered with veils of illusion, preventing us from seeing its deeper structure?

Many ancient philosophies, including Vedanta, Buddhism, and Taoism, suggest that reality is an illusion shaped by perception—a concept eerily similar to the modern understanding that space-time can be warped, manipulated, and perhaps even transcended under extreme conditions, such as inside a black hole.

Could black holes, like Rahu and Ketu, be reminders that what we see is only part of the truth, and that beyond the cosmic veil lies an even greater reality?

Conclusion: The Cosmic Veil and the Limits of Human Perception

Ancient wisdom and modern science converge on a strikingly similar idea:

- Reality is not what it appears to be.
- There exist hidden forces that shape the cosmos, yet remain beyond direct perception.
- What we see is only a fraction of what truly exists, as most of the universe remains concealed behind veils of illusion.

Whether expressed through the symbolism of Rahu and Ketu or the physics of black holes, the message is clear—there is always something beyond, something hidden, something yet to be understood.

Perhaps black holes are not cosmic endpoints, but cosmic thresholds—gates through which the universe regenerates, evolves, and reveals itself in ways we are only beginning to comprehend.

In the next chapter, we will explore how the very fabric of the universe is governed by these hidden energy flows, drawing deeper connections between ancient energy dynamics and modern cosmic structures.

3: The Cosmic Dance of Creation and Destruction

In the previous chapter, we explored how Rahu, Ketu, and black holes serve as cosmic veils, concealing deeper truths about the nature of space, time, and matter. We uncovered the parallels between Vedic cosmology and modern physics, revealing how unseen forces shape the structure of the universe. But beyond the hidden nature of reality, there is another profound truth—the universe is not static; it moves, breathes, and transforms in an eternal rhythm of creation and destruction.

This rhythm is beautifully expressed in the concept of Shiva's Tandava—the cosmic dance that governs the birth, preservation, and dissolution of existence. In Vedic tradition, Shiva's dance is not just a metaphor but a representation of the fundamental forces of the cosmos—expansion, contraction, and regeneration.

Modern astrophysics, too, describes the universe in similar terms. The Big Bang marks the universe's creation, expansion drives its evolution, and the Big Crunch (or other cyclic theories) suggests an eventual collapse, leading to possible rebirth. Could it be that what the ancients envisioned as Shiva's dance of destruction and renewal is reflected in the dynamic, ever-changing fabric of the cosmos? And if so, do black holes serve as key instruments in this celestial choreography, dissolving matter only to seed new forms of existence?

In this chapter, we step deeper into the cyclic nature of the universe, drawing parallels between Shiva's Tandava, Vedic cosmology, and modern scientific theories of cosmic expansion and contraction. Through this exploration, we may find that creation and destruction are not opposites but complementary forces, essential for the eternal dance of the cosmos.

3.1 The Tandava: Shiva's Dance of the Universe

The universe is not a still and silent entity; it is in a constant state of motion, transformation, and renewal. From the movement of galaxies to the subatomic vibrations within matter, everything in existence follows a rhythm—a cosmic dance of creation and destruction. In ancient Indian philosophy, this dance is symbolized by Shiva's Tandava, a concept that not only carries deep spiritual meaning but also finds striking parallels in modern astrophysics.

Tandava is more than just a divine performance; it is a metaphor for the dynamic nature of the cosmos. The idea that the universe is governed by cycles of birth, evolution, and dissolution is not unique to Hinduism—many cultures and philosophies have attempted to describe the pulsation of existence. However, what makes Tandava particularly fascinating is its uncanny resemblance to the behavior of the universe as described by modern science.

The Symbolic Meaning of Shiva's Tandava: Cosmic Creation and Dissolution

Shiva, in Hindu tradition, is often seen as the destroyer, but this is a misinterpretation. Shiva is not merely a force of destruction—he is the regulator of cosmic balance. His role is not to end existence but to ensure its renewal, much like the laws of thermodynamics ensure the transformation of energy rather than its disappearance.

In the Nataraja form, Shiva is depicted as:

- Dancing within a ring of fire, symbolizing the eternal cycle of the universe.
- One foot crushing Apasmara, the demon of ignorance, indicating the removal of illusion to reveal deeper cosmic

truths.

- One hand holding the drum (Damaru), representing the rhythmic beat of creation.
- Another hand holding fire, signifying the energy of dissolution and transformation.

This imagery captures the interwoven relationship between creation, sustenance, and destruction, much like how the universe expands, maintains equilibrium, and eventually contracts before a new cycle begins.

Tandava is said to contain five cosmic processes (Panchakritya):

1. Srishti (Creation) – The emergence of existence, similar to the Big Bang.

2. Sthiti (Preservation) – The stability and expansion of the cosmos.

3. Samhara (Destruction) – The eventual dissolution of forms, much like black holes consuming matter.

4. Tirobhava (Concealment) – The mystery of the unknown, much like the hidden nature of dark matter and black holes.

5. Anugraha (Grace) – The rebirth of new cycles, aligning with theories of cosmic regeneration.

Tandava is thus a symbolic representation of the fundamental processes that govern existence—from the life cycles of stars to the grand movements of galaxies.

Tandava and the Eternal Cycle of Birth, Transformation, and Destruction

The cyclic nature of existence is a core idea in both Hindu cosmology and modern physics. The ancients believed that the universe does not move in a straight line toward an ultimate end but instead oscillates between phases of creation and destruction.

The Vedic Understanding of Cyclic Time

- The concept of Yugas, Kalpas, and Pralayas suggests that time

itself is not linear but cyclical, moving through grand cosmic epochs.
- The universe is born, expands, dissolves, and is reborn—an eternal cycle.
- Pralaya (cosmic dissolution) is not a destructive force, but a state of transformation where the universe resets itself.

This view closely resembles modern scientific discussions on the nature of the cosmos.

Scientific Parallels: The Universe as a Cosmic Dance

As scientific discoveries progress, the notion of a universe that is constantly in motion, expanding, contracting, and transforming is becoming more widely accepted. The Tandava of Shiva finds its equivalent in the very laws of physics governing the cosmos.

1. Expansion and Contraction: The Universe as a Pulsating System

- The Big Bang Theory states that the universe began from an incredibly dense point and expanded outward, much like the first beat of Shiva's drum, initiating creation.
- The Expanding Universe Model (Edwin Hubble) revealed that galaxies are moving away from each other, suggesting that the universe is still expanding.
- Some theories, like the Big Bounce Hypothesis, suggest that the universe could eventually contract and restart in another Big Bang, mirroring the continuous cycles described in the Tandava.

This idea aligns perfectly with Tandava's representation of continuous creation and destruction. The universe, rather than existing in a permanent state, is constantly reshaping itself, much like Shiva's dance shaping the flow of energy.

2. Black Holes as Destructive yet Creative Forces

- Black holes, often seen as the ultimate destructive entities, actually play a key role in cosmic evolution.
- They regulate the structure of galaxies, recycle matter, and may even serve as gateways to new universes.
- This is similar to the role of Shiva, who destroys only to allow new forms of creation to emerge.

If Shiva's dance represents the continuous renewal of existence, then black holes may be one of the mechanisms through which this cosmic dance unfolds.

3. Quantum Fluctuations and the Creation of Matter

- At the smallest scales, quantum mechanics shows that particles constantly appear and disappear, much like how creation and destruction occur in Tandava.
- The uncertainty principle means that matter is never truly at rest; it is always vibrating, shifting, and transforming, just as Tandava is a dance of perpetual motion.

Could it be that Shiva's Tandava is not just a metaphor but an intuitive understanding of the fundamental forces of the universe—ones that modern physics is only now beginning to quantify?

Conclusion: The Universe as an Eternal Dance

When we observe the cosmic forces at play, it becomes clear that Tandava is more than mythology—it is a profound representation of universal dynamics.

- The universe is in constant motion, much like the dance of Shiva.
- Expansion and contraction, creation and dissolution are not separate—they are part of the same cosmic process.
- Modern physics describes the universe as a dynamic, oscillating system, an idea that mirrors the Vedic

understanding of time and cosmic cycles.

Whether through the language of spirituality or physics, the message is the same:
The universe does not move toward an absolute end—it is engaged in an eternal rhythm of existence, destruction, and rebirth.
In the next section, we will explore how Vedic cosmology's "breathing universe" aligns with scientific models of cyclic expansion and contraction, drawing deeper connections between ancient wisdom and cutting-edge astrophysics.

3.2 The Breathing Universe: Vedic Cosmology and Cyclic Existence

The universe has long been imagined as an entity that expands, contracts, and regenerates in an eternal cycle. In modern physics, some theories suggest that the cosmos does not simply expand infinitely but may one day collapse and restart in a continuous loop. This concept finds a deep parallel in Vedic cosmology, where the universe is described as undergoing periodic cycles of creation, stability, and dissolution—a process that resembles the breathing of a cosmic entity.
Ancient Hindu texts do not see time as linear; rather, they depict it as cyclical, much like the rhythms of nature. The rise and fall of civilizations, the changing of seasons, and the pulsations of the cosmos itself are all seen as reflections of a grand, repeating cosmic process. This vision, encoded in the ideas of Kalpas and Yugas, suggests that the universe has no single beginning or end but moves through alternating phases of existence.
Modern science, too, is increasingly moving toward models that describe the universe as a self-renewing system, where expansion and contraction may be fundamental to its nature. Could it be that ancient cosmologists had already grasped a concept that modern physics is only beginning to explore?

The Concept of Kalpas and Yugas: Cycles of Creation, Stability, and Dissolution

In Hindu cosmology, time is divided into vast epochs, called Kalpas and Yugas, that govern the birth, evolution, and destruction of the universe. These cycles are not random but follow a structured sequence, much like the evolution of galaxies and cosmic structures observed by astronomers today.

- A Kalpa (4.32 billion years) is one full day in the life of Brahma, the Creator, and consists of cycles of manifestation and dissolution.
- Each Kalpa is further divided into four Yugas (ages), each representing a progressive decline in cosmic order:
    - **Satya Yuga (Golden Age)** – A time of harmony and purity.
    - **Treta Yuga** – A gradual decline in righteousness.
    - **Dwapara Yuga** – Further decline, with increasing imbalance.
    - **Kali Yuga (Present Age)** – The darkest phase, marked by chaos and destruction.

After the completion of a Kalpa, the universe undergoes Pralaya (dissolution), where all matter and energy return to an unmanifest state before emerging again in the next cycle.

This view aligns strikingly well with modern cyclic universe theories, which suggest that the cosmos may go through repeated cycles of expansion and contraction.

Vedic Cosmology Describes the Universe as Breathing In and Out

In the Rig Veda and Upanishads, the universe is often compared to a breathing entity—expanding with each exhalation and contracting with each inhalation. This metaphor describes the rhythmic pulsation of existence, where periods of activity are followed by phases of dissolution.

The **Brahmanda (Cosmic Egg)** concept states that the universe undergoes a process of:

- **Srishti (Creation):** Expansion from a singular source.
- **Sthiti (Sustenance):** A period of relative stability.
- **Samhara (Dissolution):** Contraction back into an unmanifest state.

This mirrors modern scientific models that describe the potential for a contracting universe after its expansion phase, much like the act of inhaling and exhaling.

- In the Big Bang Theory, the universe is thought to have emerged from an incredibly dense state, much like the Brahmanda expanding from an unmanifest seed.
- If the universe is closed, gravity may eventually reverse the expansion, pulling everything back into a dense singularity.
- This resembles Pralaya, where the universe returns to a state of equilibrium before a new cycle begins.

Could it be that what ancient seers described as the cosmic breath is actually an intuitive understanding of the universe's expansion and contraction?

Does This Ancient Knowledge Align with Modern Cyclic Universe Models?

Modern astrophysics suggests that the universe's fate is not yet fully determined, but several theories propose a cyclic nature similar to Vedic cosmology:

1. The Big Crunch Hypothesis

- According to this theory, the universe is currently expanding due to dark energy, but if gravitational forces take over, it may eventually reverse and contract back into a singularity.

- This mirrors the concept of Pralaya, where the universe dissolves back into an unmanifest state before another cycle of creation.

2. The Big Bounce Model

- Some physicists propose that rather than ending in a singularity, the universe may undergo a Big Bounce, where each collapse triggers a new Big Bang—a continuous cycle of destruction and rebirth.
- This aligns with the idea that Kalpas and Yugas repeat endlessly, with each cycle marking a new phase of cosmic evolution.

3. Entropy Reversal and the Cyclic Universe

- In physics, entropy (the measure of disorder) tends to increase over time, which suggests that the universe is moving toward a state of greater randomness.
- However, some theories propose that entropy may reset during cosmic contraction, leading to a new low-entropy universe—a direct parallel to Pralaya resetting cosmic energy before a new creation cycle.

All of these ideas point toward a universe that does not follow a linear path to oblivion but instead regenerates itself through endless cycles—just as ancient Indian texts have described for thousands of years.

Bridging Ancient and Modern Thought: Is the Universe an Eternal Cycle?

Whether through spiritual insights or scientific exploration, the conclusion remains the same:

- The universe is not a one-time creation but an ongoing process of emergence and dissolution.
- Ancient texts describe time as a rhythmic cycle, much like the oscillating universe models in modern physics.
- The concept of Pralaya aligns with entropy reversal and energy transformation, suggesting that destruction is not an end, but a necessary phase for renewal.

If the universe does indeed breathe in and out, with alternating periods of expansion and contraction, then perhaps the wisdom of the ancients was not just mythological, but an intuitive grasp of cosmic truths that science is only now beginning to quantify.

In the next section, we will explore how the Big Crunch hypothesis and modern cosmological models compare to these ancient descriptions, further uncovering the parallels between Vedic thought and contemporary astrophysics.

3.3 The Big Crunch and the Fate of the Universe

For much of human history, civilizations have pondered the origin and fate of the universe. While ancient texts describe cosmic cycles of creation and dissolution, modern physics seeks to explain these phenomena through cosmological models and astrophysical theories. The Big Bang theory, which suggests that the universe began from an extremely dense singularity and expanded outward, is widely accepted today. However, a crucial question remains: Will this expansion continue forever, or will the universe one day contract and collapse back into itself?

One compelling theory, known as the Big Crunch hypothesis, suggests that the universe may not expand indefinitely but instead reverse its course, leading to a collapse back into a singular state. This mirrors the Vedic concept of Pralaya, where the universe dissolves into an unmanifest state before being reborn in a new cosmic cycle. Could it be that what ancient seers described as Pralaya—the periodic dissolution of the cosmos—is the same process that astrophysicists predict through the Big Crunch model?

The Big Bang and Cosmic Expansion: How Modern Physics Describes the Unfolding of the Cosmos

The modern scientific understanding of the universe's origin begins with the Big Bang theory, which states that:

- The universe started from an extremely hot, dense point called a singularity.
- It expanded rapidly, leading to the formation of subatomic particles, atoms, and eventually galaxies.
- Over billions of years, this expansion has continued, with galaxies moving farther apart, driven by an unseen force called dark energy.

The expansion of the universe is observed through redshift, a phenomenon where light from distant galaxies is stretched to longer wavelengths as they move away from us. This discovery, made by Edwin Hubble in the 1920s, provided strong evidence that space itself is expanding, carrying galaxies along with it.

However, while expansion is currently accelerating, some physicists believe that this may not be a permanent trend. If the balance of forces shifts, gravity could eventually overcome the outward momentum, pulling everything back inward—leading to the phenomenon known as the Big Crunch.

The Big Crunch Hypothesis: Will the Universe Eventually Collapse Back into a Singularity?

The Big Crunch hypothesis proposes that:

- If the density of the universe is high enough, gravity will slow down the expansion.
- Eventually, all matter and energy will stop moving outward and begin contracting.
- Galaxies, stars, and even atoms themselves will be pulled together, collapsing into a single point of infinite density—a new singularity.
- This singularity may give rise to another Big Bang, starting the cycle again.

This theory suggests that the universe is not a one-time event but a recurring phenomenon, undergoing alternating phases of expansion and contraction.

Interestingly, this idea is not exclusive to modern physics. Ancient Indian cosmology describes a strikingly similar process, where the universe undergoes Kalpas (cosmic days) and dissolves into Pralaya (cosmic dissolution) before emerging again in a new cycle.

Could This Mirror the Vedic Cycle of Pralaya?

The concept of Pralaya in Hindu cosmology is a periodic dissolution of the cosmos, which occurs at the end of each Kalpa (Brahma's day). During Pralaya:

- All physical forms dissolve, leaving only the unmanifest essence (Avyaktam).
- The universe collapses back into its fundamental state, much like a singularity in the Big Crunch model.
- A new cycle of creation follows, repeating the process endlessly.

This cyclic model of the universe closely aligns with the idea that the cosmos might undergo expansion and contraction eternally, rather than moving toward a permanent end.

Scientific and Vedic Parallels

Concept	Modern Science (Big Crunch)	Vedic Cosmology (Pralaya)
Expansion Phase	The universe expands after the Big Bang.	The universe unfolds during Brahma's day.
Contraction Phase	Gravity slows down expansion and reverses it.	Cosmic forces withdraw creation back into the unmanifest.
Singularity Collapse	The universe collapses into a dense state.	Everything merges back into Avyaktam (unmanifest state).
Rebirth Cycle	The singularity may trigger another Big Bang.	A new cycle of creation begins after Pralaya.

The remarkable similarity between these two perspectives suggests that ancient thinkers may have intuited the oscillating nature of the cosmos long before scientific instruments could measure it.

Challenges to the Big Crunch Theory and Alternative Views

While the Big Crunch hypothesis aligns with the idea of cosmic cycles, modern observations suggest that the universe's expansion is accelerating due to dark energy. If this acceleration continues indefinitely, then the universe may:

- Expand forever in a "Big Freeze," where it becomes cold, dark, and lifeless as galaxies drift apart.
- Tear apart in a Big Rip, where the expansion of space eventually destroys all matter at a fundamental level.

However, some physicists believe that dark energy itself could change over time, possibly leading to a contraction phase in the distant future.

Additionally, the Big Bounce theory suggests that even if a Big Crunch occurs, the universe does not end permanently but instead rebounds into another Big Bang, leading to a new expansion.

This idea is particularly intriguing because it mirrors the Vedic notion of Kalpas and Pralaya as part of an eternal cycle.

Conclusion: The Eternal Rhythm of the Universe

Whether through the language of science or the wisdom of ancient philosophy, the universe appears to be governed by cycles of emergence and dissolution.

- The Big Bang and cosmic expansion parallel the Vedic concept of Srishti (creation).
- The Big Crunch hypothesis mirrors the idea of Pralaya, where all matter returns to an unmanifest state.
- If the universe follows a cyclic pattern, then the scientific concept of the oscillating universe and the Vedic idea of Kalpas align in profound ways.

Could it be that our universe is simply one iteration in an infinite cosmic cycle? If so, both ancient wisdom and modern physics may be pointing to the same fundamental truth—that the universe is neither eternal nor finite, but an ever-renewing dance of energy and transformation.

In the next section, we will explore how black holes, as cosmic regulators, might play a role in this grand cycle, influencing the destruction and rebirth of cosmic structures.

3.4 Black Holes and the Cosmic Rhythm

In the grand design of the universe, destruction is not an end, but a transformation. Just as Shiva's Tandava symbolizes the cyclical dance of creation, preservation, and dissolution, modern astrophysics suggests that black holes are not mere cosmic voids but active participants in shaping and renewing the cosmos.

Black holes are the most enigmatic objects in the universe. They consume everything that crosses their event horizon, yet they may also play a crucial role in redistributing matter and energy, much like cosmic recyclers. This paradox of destruction leading to regeneration echoes the role of Shiva as the cosmic dancer, whose destructive energy clears the way for new creation.

Could black holes be the instruments of the universe's eternal rhythm—the destroyers that ensure cosmic renewal?

Are Black Holes the "Destroyers" in the Cosmic Dance?

At first glance, black holes seem like devourers of existence, consuming everything, from light to entire stars. Their immense gravitational pull traps matter in such a way that nothing, not even information, was once thought to escape. This led to the view that black holes represent an absolute end—like a cosmic death.

However, emerging research suggests that black holes do not just destroy; they also transform and redistribute energy in unexpected ways. Much like Shiva's Tandava, which dissolves old structures to give rise to new ones, black holes:

- Consume matter but also release high-energy jets that influence star formation.
- Distort space-time, altering the movement of galaxies and cosmic structures.
- May serve as transition points, where matter undergoes transformation and re-emerges elsewhere in the universe.

This challenges the idea of black holes as mere "endpoints" and instead presents them as fundamental players in the cosmic cycle of renewal.

How Black Holes Reshape Galaxies and Cosmic Structures

Black holes, particularly supermassive black holes, are now recognized as key architects of galactic evolution. These massive entities—millions to billions of times the mass of our Sun—reside at the centers of most galaxies, including the Milky Way.

Ways Black Holes Influence the Cosmos:

1. Regulating Star Formation

- Black holes generate powerful outflows of energy, which can either trigger the birth of new stars or shut down star formation.
- This mirrors Shiva's role in both destruction and regeneration—where one cycle ends, another begins.

2. Maintaining Galactic Structure

- Without the influence of black holes, galaxies may not hold their shape or form in the way we observe.
- Their immense gravity stabilizes galactic centers, ensuring a balance between expansion and collapse.

3. Cosmic Recycling of Matter

- As black holes consume matter, they release enormous bursts of energy, enriching their surroundings with heavy elements.
- This fuels the formation of new celestial bodies, ensuring that destruction leads to creation.

This cycle is eerily similar to how Shiva's dance destroys the old but also creates the conditions for renewal. If Shiva's Tandava maintains the balance of existence, could black holes be the cosmic equivalents, ensuring the universe remains dynamic and ever-evolving?

Could Black Holes Lead to the Next Phase of Cosmic Rebirth?

While black holes have long been seen as final destinations for matter, new theories suggest that they might be gateways to something more profound.

1. White Holes: The Other Side of Black Holes?

- Some physicists propose that black holes could be linked to white holes—hypothetical objects that expel matter instead of absorbing it.
- This raises the possibility that what falls into a black hole may re-emerge elsewhere in the universe—or even in a different universe.
- If true, black holes may not be cosmic endpoints but transition points, allowing matter to be recycled into new realities.

2. The Role of Black Holes in the Big Bounce Theory

- The Big Bounce theory suggests that instead of a permanent collapse, the universe contracts into a highly dense state before rebounding into another expansion phase.
- If black holes play a role in gathering and concentrating matter, they could serve as seeds for the next cosmic cycle.
- This mirrors the idea that Shiva's destruction in Tandava does not signify an end, but a necessary transformation for new creation to emerge.

3. Black Holes as Pralaya in the Cosmic Cycle

- In Vedic cosmology, Pralaya (cosmic dissolution) is not an absolute end but a phase where the universe rests before being reborn.
- Black holes seem to perform a similar function, taking in matter, breaking it down into its most fundamental state, and

possibly redistributing it across the cosmos.

Could it be that black holes are the cosmic agents of Pralaya, ensuring that matter is never truly lost but instead transformed into the foundation for a new universe?

Conclusion: Black Holes as the Cosmic Dancers of Renewal

Much like Shiva's Tandava, black holes are both destroyers and creators—they dissolve matter only to allow new cosmic structures to emerge.

- They consume but also create.
- They collapse matter but release energy.
- They appear to end existence but may serve as portals to new beginnings.

This profound duality bridges the ancient wisdom of Vedic cosmology with modern astrophysical discoveries, reinforcing the idea that destruction and creation are not opposites, but necessary components of the universe's eternal cycle.

In the next section, we will explore whether the entire universe itself functions like a giant black hole—an entity that collapses and re-emerges, mirroring the eternal cosmic dance Shiva performs across time and space.

3.5 The Universal Balance: Destruction as a Necessity for Creation

In both ancient wisdom and modern science, destruction is not viewed as an end, but as an essential process for transformation and renewal. Whether in the form of Shiva's Tandava, the cycles of Pralaya in Vedic cosmology, or the behavior of black holes in astrophysics, we see a recurring theme—existence is constantly evolving through cycles of creation and dissolution.

From the death of stars to the collapse of galaxies, destruction fuels the birth of new cosmic structures. Even at the subatomic level, particles interact, decay, and reform, ensuring that matter and energy never disappear, but only change forms. This principle aligns with the law of conservation of energy in physics and the philosophy of cyclic existence found in spiritual traditions.

Could it be that the universe itself is designed to reset, ensuring an eternal balance between formation and annihilation?

Destruction as Transformation: A Core Concept in Physics and Spirituality

In Hindu philosophy, destruction is often misunderstood. Shiva, the destroyer, is not a force of annihilation, but a cosmic transformer. The destruction he represents is not chaos but a necessary force that removes the old, making way for the new.

This idea is mirrored in modern physics, which tells us that matter and energy cannot be destroyed—only transformed.

Key Scientific and Spiritual Parallels:

Concept	Spiritual Interpretation (Vedic Cosmology)	Scientific Interpretation (Physics & Cosmology)
Destruction as Renewal	Shiva's Tandava destroys outdated forms to create new realities.	Supernovas and black holes break down matter to form new celestial bodies.
Matter is never lost	Energy in the universe is eternal, cycling through creation and dissolution.	The First Law of Thermodynamics states that energy can neither be created nor destroyed.
Cyclic Nature of Time	The universe moves through Kalpas (creation cycles) and Pralaya (dissolution).	The cyclic universe model suggests repeated Big Bangs and Big Crunches.
Chaos Leading to Order	Shiva's dance appears destructive but is part of a higher cosmic order.	The universe evolves through entropy and structure formation.

From the destruction of a dying star to the recycling of elements into new planetary systems, we see this principle of renewal in action everywhere.

How Entropy, Black Holes, and Cosmic Cycles Drive Evolution

A key concept in physics that governs the direction of change in the universe is entropy—the measure of disorder or randomness in a system.

1. Entropy and the Arrow of Time:

- The Second Law of Thermodynamics states that entropy always increases, meaning that systems move toward greater

disorder over time.
- However, in the grand scale of the universe, entropy does not mean destruction—it leads to reorganization and new structures.
- Just as Shiva's dance appears chaotic but follows an underlying rhythm, entropy ensures that matter and energy keep transitioning into new forms.

2. Black Holes as Cosmic Recyclers:

- When a massive star dies, it collapses into a black hole, seemingly destroying everything.
- However, the black hole's immense gravity creates energy jets, feeding new star formation in its surrounding galaxy.
- This mirrors the role of Shiva in dissolving the old to allow new realities to emerge.

3. Cosmic Cycles and the Evolution of the Universe:

- The universe started as a highly ordered, low-entropy state in the Big Bang.
- Over billions of years, stars formed, galaxies evolved, and entropy increased.
- At some point, the universe may reverse this process, either collapsing in a Big Crunch or resetting itself.

This leads to the ultimate question:

Is the Universe Designed to Reset Itself Indefinitely?

Many scientific theories suggest that the universe is not moving toward a single, permanent state but is instead engaged in an endless cycle of transformation.

1. The Cyclic Universe Model

- Some physicists propose that after reaching maximum expansion, the universe will collapse back into a singularity, triggering another Big Bang.
- This aligns with the idea of Kalpas and Pralaya, where each cosmic cycle ends in dissolution, only to begin anew.

2. The Big Bounce Theory

- Instead of a permanent collapse, the universe may bounce back after contraction, leading to another expansion phase.
- This is similar to Shiva's eternal dance, where destruction is immediately followed by renewal.

3. The Multiverse Hypothesis

- Some suggest that black holes could be gateways to new universes, meaning that even if our universe "ends," new realities emerge from its remnants.
- This resonates with the Vedic concept of multiple cosmic realms, each undergoing its own cycles of creation and destruction.

If these models hold true, then the universe is not a one-time occurrence, but part of an eternal, self-renewing system.

Conclusion: The Eternal Balance of Creation and Dissolution

Destruction is often feared because it represents an end—but in both spiritual philosophy and scientific thought, destruction is not an endpoint but a transition.

- Black holes, entropy, and cosmic cycles ensure that the universe remains in a state of constant renewal.
- The death of stars leads to the formation of new ones, just as

galaxies are reshaped by the forces of cosmic evolution.
- The parallels between modern physics and Vedic cosmology suggest that the universe is structured to maintain an eternal balance.

If the universe truly resets itself indefinitely, then the ancient wisdom of Shiva's dance, the cycles of Kalpas, and modern theories of cosmic evolution may all be pointing toward the same fundamental reality—The cosmos is not a singular event, but an eternal rhythm—one where destruction is not an end, but a necessity for creation.

4: Pralaya – The Science of Universal Dissolution

In the previous chapter, we explored the cosmic balance between destruction and creation, revealing how black holes, entropy, and cosmic cycles ensure that existence is constantly evolving. But what happens when everything dissolves? Is destruction truly an end, or is it part of a larger rhythm of renewal?

Ancient Indian cosmology introduces Pralaya, a concept that describes the dissolution of the universe, marking the transition between one cosmic cycle and the next. Unlike apocalyptic visions of total annihilation, Pralaya is not an absolute end but a restructuring process—a phase where matter dissolves into its fundamental essence before emerging again in a new form.

Surprisingly, this idea is not far from modern astrophysical theories. The death of stars, the expansion and collapse of galaxies, and even the potential fate of the entire universe bear remarkable similarities to the four types of Pralaya described in Vedic texts. Whether through the gradual decay of systems, the fiery destruction of celestial bodies, or the ultimate collapse of the cosmos into a singularity, dissolution appears to be an essential function of the universe.

Could it be that what the ancients called Pralaya is the same cosmic process that astrophysicists study today—the slow breakdown and transformation of the universe? In this chapter, we explore the different types of Pralaya, their scientific parallels, and the profound question: Does the universe truly end, or is it destined to be reborn eternally?

4.1 Understanding Pralaya: The Four Types of Cosmic Dissolution

The universe, both in ancient cosmology and modern physics, is never static—it constantly transforms, decays, and regenerates. In Vedic thought, this cycle is understood through Pralaya, the dissolution of cosmic structures at different scales and timelines. Unlike the idea of an apocalyptic end, Pralaya is a natural phase of cosmic evolution, much like how stars die, galaxies collide, and matter is continuously restructured in astrophysical processes.

Hindu cosmology categorizes Pralaya into four types, each describing different levels of dissolution—from minor transformations to the complete collapse of the cosmos. What is striking is how these descriptions align with modern scientific theories about the fate of celestial bodies, the death of stars, and even the potential end of the universe itself.

The Four Types of Pralaya

1. Naimittika Pralaya – The Cyclic Dissolution of Worlds

(The End of an Age, but Not the Universe)

Naimittika Pralaya occurs at the end of a Kalpa (Brahma's day), which lasts 4.32 billion years. It is a temporary dissolution, where certain cosmic systems are destroyed, but the larger universe remains intact.

Scientific Parallels:

- The death of stars in supernova explosions, where individual celestial bodies collapse while the galaxy persists.
- The expansion of the Sun into a red giant, consuming Earth, yet leaving the larger Milky Way unaffected.

- Galactic collisions, where some stars are destroyed, but new ones are born in the process.

In this phase, destruction is localized, much like how certain regions of the universe undergo change while others continue to evolve.

2. Prakritika Pralaya – The Dissolution of the Physical Universe

(The Breakdown of Matter and Space-Time Itself)
Prakritika Pralaya is more profound—it occurs when the entire material universe dissolves into its fundamental elements, leaving only unmanifest cosmic energy (Mula Prakriti). This is believed to happen when Brahma's life of 100 years ends, meaning after 311 trillion years.

Scientific Parallels:

- The Heat Death of the Universe (Big Freeze), where the universe expands until all energy is evenly distributed, leading to a state of no usable energy and no possibility of life.
- Vacuum Decay, a theoretical quantum physics scenario where a small quantum fluctuation could cause space-time to collapse, wiping out all matter instantly.
- Black hole mergers and singularity formation, where matter is completely broken down into pure energy.

In this stage, the laws of physics as we know them cease to operate, resembling the scientific concept that at the end of the universe, all structured matter will break down into its simplest, most disordered state.

3. Nitya Pralaya – Continuous Micro-Dissolutions in the Cosmos

(The Perpetual Transformation of Reality at Every Moment)
Nitya Pralaya is not a grand cosmic event but a constant process of decay and renewal that happens every moment. Every day, planets shift, stars lose energy, and life forms die. This dissolution is not an end but part of the ongoing cycle of impermanence.

Scientific Parallels:

- The death of individual organisms, mirroring the natural cycle of life and death.
- Radioactive decay, where atoms constantly transform into different elements over time.
- The Hawking Radiation hypothesis, where black holes gradually evaporate, losing their energy bit by bit over trillions of years.
- The gradual decay of orbits and planetary systems, which eventually destabilize over millions or billions of years.

This type of Pralaya represents the fundamental truth of impermanence—nothing remains unchanged, even on the smallest scale.

4. Maha Pralaya – The Ultimate Cosmic Dissolution

(The Absolute End and Rebirth of Everything)
Maha Pralaya is the final and total dissolution—the moment when all universes, dimensions, and even the fabric of space-time collapse into the unmanifest. It is the end of all existence before a new cycle begins.

Scientific Parallels:

- The Big Crunch Theory, where the universe's expansion reverses, causing everything to collapse back into a singularity.
- The Big Bounce Model, which suggests that after the universe collapses, it will expand again in a new Big Bang—aligning with the Vedic concept of the next cosmic cycle.
- The ultimate singularity of black holes, where all known physics breaks down, leading to unknown possibilities.

Maha Pralaya aligns closely with the modern idea that the universe is not a one-time event, but a repeating cycle of creation and destruction.

How the Four Types of Pralaya Map onto Scientific Models

Type of Pralaya	Vedic Description	Scientific Parallel
Naimittika Pralaya	Partial destruction at the end of a cosmic age.	Death of stars, planetary destruction, galactic collisions.
Prakritika Pralaya	Total dissolution of the physical universe into primal energy.	Heat death, vacuum decay, breakdown of fundamental particles.
Nitya Pralaya	Constant small-scale dissolution in daily cycles.	Radioactive decay, Hawking radiation, biological death.
Maha Pralaya	The absolute end of all space, time, and matter.	Big Crunch, Big Bounce, black hole singularities.

The striking similarities between these ancient descriptions and modern cosmological theories suggest that the concept of Pralaya is not merely symbolic but a profound intuitive understanding of cosmic evolution.

Conclusion: Is Pralaya a Scientific Reality?

Pralaya, as described in Vedic cosmology, does not represent a supernatural destruction but a natural phase in the endless cycle of the cosmos.

- Every level of existence—from stars to entire universes—undergoes periodic transformation, much like the four types of Pralaya.
- Modern astrophysics confirms that celestial bodies, galaxies, and even the entire universe may one day face total dissolution.
- Whether it is stellar death, entropy-driven collapse, or a complete cosmic reset, the parallels between Pralaya and scientific theories are undeniable.

The universe, it seems, is designed to break down and rebuild itself eternally, maintaining a balance between creation and destruction.
In the next section, we will explore how specific Vedic descriptions of cosmic fire, elemental dissolution, and energy transformation align with real astrophysical phenomena, including black holes, supernovae, and quantum field fluctuations.

4.2 Naimittika Pralaya: The Temporary Dissolution of Worlds

In the grand cosmic timeline, not all destruction is absolute. Some dissolutions are temporary, affecting only specific celestial bodies while the larger cosmic framework continues. This is precisely what Naimittika Pralaya describes—a periodic dissolution that occurs at the end of a Kalpa (Brahma's day), lasting 4.32 billion years. It is a moment of transition, where certain cosmic structures—such as planetary systems or stars—are erased, yet the universe itself remains intact.

In modern astrophysics, this concept finds striking parallels in the life cycles of stars, planetary destruction, and transient cosmic events. From the Sun's eventual transformation into a red giant, engulfing the inner planets, to the violent death of massive stars in supernova explosions and gamma-ray bursts, Naimittika Pralaya describes the inevitable impermanence of individual celestial bodies within the grander structure of the cosmos.

The Cyclic Destruction of Planets and Stars within a Larger Cosmic Timeline

Ancient texts suggest that at the end of each Kalpa, Lord Vishnu, in his cosmic form, withdraws the material world into himself, causing the temporary destruction of stars and planetary systems. However, this is not the final end of existence—it is merely a phase in an ongoing cycle of creation and destruction.

This perspective aligns with modern astrophysical observations that confirm:

- Stars are not eternal—they are born, live, and eventually die, either fading away or exploding violently.
- Planetary systems undergo catastrophic changes over time due to stellar evolution, collisions, and cosmic impacts.
- Galaxies experience large-scale restructuring, with entire regions undergoing temporary dissolution.

This type of localized, yet profound, cosmic destruction is observable in the universe today.

How the Sun's Expansion and Eventual Engulfing of Earth Mirrors Naimittika Pralaya

One of the most well-understood examples of Naimittika Pralaya in action is the fate of our own Solar System.

- The Sun, like all stars, follows a predictable life cycle, currently in its main sequence phase.

- However, in about 5 billion years, it will run out of hydrogen fuel and enter the red giant phase.
- As the Sun expands, its outer layers will engulf Mercury, Venus, and possibly Earth.
- This transformation will mark the destruction of our planetary environment—yet the rest of the universe will continue unaffected.

This matches the concept of Naimittika Pralaya, where individual cosmic systems undergo dissolution at the end of an era, but the larger cosmos remains intact.

Furthermore, this process is not unique to our Sun—countless stars in the universe undergo similar phases, leading to the destruction of their surrounding planets.

Supernovas and Gamma-Ray Bursts: Real Astrophysical Phenomena That Erase Worlds

While the Sun will expand and fade away relatively peacefully, massive stars undergo a far more violent form of dissolution—supernova explosions.

Supernovas occur when:

- A star runs out of nuclear fuel and can no longer resist gravitational collapse.
- Its core implodes, generating an immense shockwave.
- The outer layers explode outward, releasing more energy in seconds than the Sun will in its entire lifetime.

These explosions are among the most destructive forces in the universe, capable of erasing entire planetary systems.

Gamma-Ray Bursts: The Most Violent Cosmic Events

- In some cases, supernovae give rise to gamma-ray bursts (GRBs), the most energetic events ever observed.
- A single GRB can release as much energy in seconds as the Sun will emit over billions of years.
- If such an event were to occur near Earth, it could strip away our atmosphere and trigger mass extinction.

Both supernovae and GRBs serve as real-world analogs of Naimittika Pralaya, where individual worlds and star systems undergo destruction as part of the natural cycle of cosmic evolution.

Conclusion: The Recurring Dissolution of Cosmic Structures

Naimittika Pralaya is not an abstract philosophical idea—it is an observable phenomenon in the death and rebirth of stars, planetary destruction, and transient cosmic catastrophes.

- The expansion and eventual collapse of the Sun aligns with the Vedic description of temporary dissolution at the end of a Kalpa.
- Supernovae and gamma-ray bursts actively reshape the universe, wiping out star systems while seeding new cosmic formations.
- This cycle of destruction is essential to cosmic renewal, ensuring that matter and energy are constantly repurposed into new stars, planets, and galaxies.

Naimittika Pralaya teaches us that no planetary system is eternal, and no star shines forever. While these events may seem destructive, they pave the way for new cosmic creation.

In the next section, we will explore Prakritika Pralaya—the total dissolution of all physical matter, where the very fabric of reality breaks down into its primordial essence, aligning with scientific theories of the universe's ultimate fate.

4.3 Prakritika Pralaya: The Dissolution of the Physical Universe

While Naimittika Pralaya describes the temporary dissolution of worlds, Prakritika Pralaya goes far deeper—it is the complete breakdown of all material existence. In Hindu cosmology, Prakritika Pralaya occurs at the end of Brahma's lifespan (311 trillion years), when the entire universe collapses into its primordial, unmanifest state (Mula Prakriti). Unlike localized destruction, this is an irreversible cosmic dissolution where time, space, and matter cease to exist.

Modern cosmology proposes similar end-of-the-universe scenarios, such as the Heat Death (Big Freeze), Vacuum Decay, and the collapse of the quantum vacuum. These scientific theories describe the gradual or sudden decay of the entire universe, leading to a state of absolute stillness or non-existence—strikingly similar to Prakritika Pralaya.

Could it be that this ancient description of ultimate dissolution aligns with modern physics' predictions for the fate of the universe?

The Destruction of All Material Existence and Its Comparison with Heat Death

Ancient Hindu texts state that in Prakritika Pralaya, all physical matter dissolves, and the universe reverts to its unmanifest state. This concept closely resembles the modern Heat Death scenario, also known as the Big Freeze.

The Big Freeze: A Universe Without Energy or Motion

The Second Law of Thermodynamics states that entropy (disorder) always increases over time. In other words, the universe is constantly moving toward a state of maximum entropy, where energy is evenly spread and no useful work can be done.

Scientists predict that if this process continues indefinitely:

- Stars will burn out, galaxies will fade, and all matter will decay.
- The universe will reach thermal equilibrium, with no heat differences left to drive motion.
- At this point, no new stars will form, and the universe will be cold, dark, and lifeless.

This final state—where nothing moves, nothing evolves, and the cosmos is devoid of energy—bears a striking resemblance to Prakritika Pralaya, where all creation dissolves back into its fundamental state of non-existence.

In Hindu cosmology, the unmanifest state is not an absolute end but a transition phase before the next creation cycle. This concept aligns with modern physics' speculation that even after Heat Death, quantum fluctuations or new cosmic events might restart the cycle.

Elemental Dissolution: How Vedic Cosmology Relates to Thermodynamics and the Fate of Matter

In Vedic cosmology, Prakritika Pralaya is described as the dissolution of all five elements (Pancha Mahabhutas):

- Earth (Prithvi) dissolves into Water (Apas).
- Water dissolves into Fire (Agni).
- Fire dissolves into Air (Vayu).
- Air dissolves into Space (Akasha).
- Space merges into the unmanifest cosmic essence (Mula Prakriti).

This step-by-step breakdown resembles the scientific idea of the gradual decay of all matter, ultimately leaving behind only fundamental energy.

Scientific Parallels to Elemental Dissolution

- **Earth dissolving into Water** – Solid matter breaking down into fundamental atomic and subatomic particles.

- **Water dissolving into Fire** – Particles reaching extreme energy states, eventually dissipating.
- **Fire dissolving into Air** – The remaining particles losing mass and transforming into pure energy.
- **Air dissolving into Space** – The final decay of energy into an empty, structureless quantum vacuum.
- **Space dissolving into the unmanifest** – The complete loss of structured space-time, leaving behind an undefined quantum state.

This breakdown mirrors modern theories about how matter will eventually decay into energy and then fade into the vacuum of space.

Does Prakritika Pralaya Align with Theories of Vacuum Decay and Quantum Instability?

In addition to Heat Death, some physicists believe that the universe could end far more suddenly through a process known as vacuum decay.

1. The Vacuum Decay Hypothesis: A Cosmic Death Wave

- Our universe exists in what is called a false vacuum—a stable energy state, but not the most stable possible.
- If a quantum fluctuation were to trigger a phase transition, it could create a true vacuum bubble where the laws of physics are different.
- This bubble would expand at the speed of light, annihilating everything in its path.

This aligns with Prakritika Pralaya, where the universe collapses back into its original, unmanifest state—comparable to a transition from a false vacuum to a true vacuum.

2. The Collapse of Space-Time

Some physicists suggest that at some point, space-time itself may become unstable, leading to:

- The breakdown of fundamental forces.
- The collapse of quantum fields.
- The loss of all known physical structures, reducing everything to a formless void.

This, again, mirrors the Vedic description of Prakritika Pralaya, where space itself dissolves into the unmanifest.

Conclusion: Is the Ultimate Dissolution of the Universe Inevitable?

Prakritika Pralaya describes not just the destruction of planets or stars, but the dissolution of all material existence into its primordial state.

Modern physics predicts similar scenarios:

- The Big Freeze could lead to a cold, motionless universe—matching the unmanifest state of Pralaya.
- Vacuum Decay could instantaneously erase all cosmic structures—similar to the abrupt dissolution of the five elements.
- Quantum instability could cause the collapse of space-time, reflecting the idea that the universe dissolves into pure potential energy.

The key takeaway is that whether gradually (through entropy) or suddenly (through vacuum collapse), science acknowledges that the universe may not last forever in its current form.

However, Hindu cosmology suggests that this dissolution is not final—it is merely a transition phase before the next cosmic cycle begins.

In the next section, we will explore Nitya Pralaya—the continuous transformation of existence, and how black holes, entropy, and quantum fluctuations ensure that nothing remains unchanged, even in the vastness of the cosmos.

4.4 Nitya Pralaya: The Continuous Transformation of Reality

While Naimittika Pralaya and Prakritika Pralaya describe periodic and total dissolutions of the cosmos, Nitya Pralaya presents a more subtle and ever-present reality—the ongoing transformation that occurs at every moment. Unlike the grand cosmic events that reshape galaxies and universes, Nitya Pralaya refers to the continuous cycle of creation and destruction at all levels of existence.

From the birth and death of cells within our bodies to the gradual decay of celestial structures, from the collapse of civilizations to the evaporation of black holes, nothing in the universe remains static. Everything is in a state of change, undergoing a process of dissolution and renewal.

This aligns perfectly with modern physics, which shows that from subatomic particles to the largest cosmic structures, impermanence is the only constant.

The Daily Cycles of Creation and Destruction

In Vedic cosmology, Nitya Pralaya is not a catastrophic event but an ongoing process that takes place every moment at multiple levels:

1. Biological Cycles:

- Cells in our bodies die and regenerate constantly—every second, millions of cells perish and are replaced.
- The human body itself is entirely renewed every few years, illustrating the continuous cycle of birth and decay.

2. Planetary and Geological Changes:

- Mountains erode, rivers change course, and ecosystems evolve.
- Entire species go extinct, making way for new life forms.

3. Civilizational and Societal Shifts:

- Empires rise and fall, cultures evolve, and societal structures are in constant flux.
- Technological advancements replace outdated ways of living, just as old systems collapse to make way for new ones.

This principle of continuous transformation mirrors the laws of physics, thermodynamics, and entropy—where change is inevitable and nothing remains unchanged forever.

How Black Holes Gradually "Consume" Their Galaxies Over Time

On a cosmic scale, Nitya Pralaya manifests in the slow but inevitable transformation of galaxies, stars, and black holes. While dramatic events like supernovae and gamma-ray bursts reshape the universe suddenly, most cosmic changes happen gradually over millions or billions of years.

1. The Galactic Recycling Process:

- Stars form from clouds of gas and dust, burn for billions of years, then die as white dwarfs, neutron stars, or black holes.
- Black holes pull in matter slowly over time, altering the structure of their host galaxies.
- The Milky Way, for example, is losing stars while new ones are forming—demonstrating a balance of destruction and creation.

2. The Eventual Evaporation of Black Holes:

- Even black holes, the most seemingly permanent entities, are not eternal.

- Hawking radiation predicts that over trillions of years, black holes will slowly lose energy and eventually vanish.
- This suggests that even the most massive objects in the universe undergo Nitya Pralaya, dissolving into nothingness over immense timescales.

Black holes thus embody the idea of continuous dissolution, much like Shiva's Tandava—where destruction is an integral part of the universe's ongoing dance of transformation.

The Impermanence of All Things: From Subatomic Particles to Entire Civilizations

The principle of Nitya Pralaya is deeply ingrained in the nature of reality, affecting everything from the smallest quantum particles to entire civilizations.

1. The Quantum Level: Constant Particle Decay and Rebirth

- At the smallest scales, particles continuously appear and disappear in quantum fluctuations.
- The Uncertainty Principle shows that matter is never truly stable; instead, it exists in a dynamic state of constant change.
- Radioactive elements decay over time, breaking down into new forms.

This confirms that Nitya Pralaya is occurring even at the most fundamental levels of physics.

2. The Rise and Fall of Civilizations

- No civilization, no empire, no nation lasts forever.
- Ancient empires like Rome, Mesopotamia, and the Indus Valley once flourished but eventually collapsed.
- Modern societies, despite their advancements, are also subject

to cycles of transformation, adaptation, and dissolution.

Nitya Pralaya reminds us that nothing is permanent, and all things must eventually change.

Conclusion: Transformation is the Essence of Reality

Unlike the sudden destruction of Naimittika Pralaya or the absolute dissolution of Prakritika Pralaya, Nitya Pralaya is an eternal process—the slow, ongoing transformation of everything in existence.

- Life and death, formation and decay, the expansion and contraction of galaxies—all are part of the ever-moving cycle of cosmic change.
- Black holes consuming galaxies, civilizations collapsing and reforming, and quantum fluctuations in empty space all illustrate the universality of this process.
- Nothing is static. Everything is in motion, everything is changing, and nothing lasts forever.

The ancient sages who described Nitya Pralaya were not just offering a philosophical insight—they were describing the fundamental nature of the universe, which modern science now confirms.

In the next section, we will explore Maha Pralaya—the ultimate dissolution of everything, where even time and space cease to exist, mirroring the theories of the Big Crunch, the singularity, and the cyclical rebirth of the cosmos.

4.5 Maha Pralaya: The Ultimate Cosmic Reset

If Nitya Pralaya represents the continuous transformation of reality, and Prakritika Pralaya signifies the gradual breakdown of the physical universe, then Maha Pralaya is the final, ultimate dissolution of everything—the complete reset of the cosmos.

Maha Pralaya is not just the destruction of stars, galaxies, or matter—it is the dissolution of time, space, and the very fabric of existence. Hindu cosmology describes it as the moment when all of creation collapses back into the unmanifest (Mula Prakriti), marking the end of one universal cycle and setting the stage for the next.

This idea is remarkably similar to modern cosmological theories about the possible fate of the universe, particularly the Big Crunch, the Big Bounce, and the possibility of a multiverse transition. Could it be that what the ancients described as Maha Pralaya is the very process by which universes die and are reborn?

What Happens When the Entire Universe Collapses?

According to Hindu cosmology, Maha Pralaya occurs at the end of Brahma's 100-year lifespan, an incomprehensibly vast period of 311 trillion years. At this point:

- The three worlds (physical, astral, and causal) dissolve.
- The five elements—Earth, Water, Fire, Air, and Space—break down into pure energy.
- Even time itself ceases to exist, merging into the unmanifest source.

This describes a universe that collapses in on itself, erasing all form, structure, and distinction. Modern physics presents similar scenarios for the end of the cosmos, particularly through gravitational collapse and singularity formation.

The Parallels Between Black Hole Singularities and Maha Pralaya

Black holes provide a fascinating parallel to the concept of Maha Pralaya.

1. A Singular Point of No Return:

- A black hole absorbs everything—matter, light, and even

space-time itself.

- Once inside the event horizon, nothing can escape, and all information is seemingly lost.
- This is similar to Maha Pralaya, where everything collapses into an unmanifest singularity.

2. Complete Loss of Information:

- According to classical physics, black holes erase information.
- Quantum physics suggests that information may be preserved in unknown ways, just as Hindu cosmology suggests that creation is never truly lost but remains in a dormant state before the next cycle.

3. The Ultimate Boundary of Physics:

- Inside a black hole, all known laws of physics break down.
- At Maha Pralaya, even space and time cease to exist, meaning that no physical law applies.
- Both suggest a state beyond human comprehension, where existence itself is reset.

If black holes are miniature versions of cosmic dissolution, could the entire universe one day collapse into a final, all-encompassing singularity—mirroring Maha Pralaya?

Is Maha Pralaya the Equivalent of the Big Crunch, Big Bounce, or a Transition to a New Universe?

Scientists today propose several possible fates of the universe, many of which align with the Vedic concept of Maha Pralaya.

1. The Big Crunch: The Universe's Collapse into a Singularity

- The Big Crunch suggests that if gravity overcomes the force of cosmic expansion, the universe will contract.

- Over billions of years, galaxies would stop moving apart and instead collapse inward.
- Eventually, everything would shrink back into a singularity—a state of infinite density and energy.
- This is exactly how Maha Pralaya is described: a cosmic contraction back into its source.

2. The Big Bounce: A New Universe After Dissolution

- Some physicists believe that after the universe collapses, a new Big Bang could occur, starting the cycle again.
- This aligns perfectly with the Vedic model of Kalpas and Maha Pralaya, where dissolution is followed by a new creation cycle.
- If our universe is cyclic, then Maha Pralaya is not a permanent end but a transition into the next cosmic age.

3. Multiverse Transitions: The Universe Merging into a Higher Reality

- Some modern theories suggest that our universe may be part of a larger multiverse.
- When our universe ends, it may collapse into a higher-dimensional space, becoming part of a larger cosmic structure.
- This resembles the Vedic belief that Maha Pralaya is not an ultimate end, but a return to a greater unmanifest state before another creation emerges.

In all these scenarios, the idea of an endless cycle of cosmic dissolution and rebirth is central, just as described in Hindu cosmology thousands of years ago.

Conclusion: Is the Universe Designed to Reset?

Maha Pralaya represents the ultimate cosmic dissolution, where even time and space collapse into the unmanifest. Modern science offers strikingly similar possibilities:

- The Big Crunch aligns with the idea of the universe contracting into a singularity.
- The Big Bounce mirrors the cyclical model of Pralaya leading to new creation.
- Black holes serve as mini-dissolutions, reflecting the grand cosmic reset of Maha Pralaya.

If these theories are correct, then the universe is not moving toward a permanent end, but is part of an eternal, repeating cycle. Just as Shiva's Tandava symbolizes destruction and renewal, Maha Pralaya ensures that the cosmic dance continues forever.

In the next section, we will explore the Vedic concept of cosmic firestorms, elemental dissolution, and energy transformation—analyzing whether these descriptions align with real astrophysical phenomena, such as Hawking radiation, high-energy cosmic events, and the evaporation of black holes.

4.6 The Firestorm of Cosmic Dissolution

In many ancient traditions, the end of time is often associated with a great cosmic fire that consumes creation, leaving behind nothing but an unmanifest void. Hindu cosmology describes this fiery dissolution as a critical phase of Pralaya, where the universe is engulfed in flames before ultimately dissolving into its primordial state.

Modern physics, too, explores the idea that the universe may end in a final burst of energy, much like the Vedic description. The concept of Hawking radiation, which predicts that black holes gradually evaporate by emitting energy, presents an intriguing parallel to the idea of cosmic dissolution through fire. Furthermore, some cosmological models suggest that high-energy processes, such as phase transitions or vacuum decay, could bring about a sudden, universe-wide energy release—an event eerily similar to the Vedic firestorm.

Could it be that the firestorm of Pralaya represents a scientific phenomenon—the final energy burst that marks the dissolution of the cosmos?

The Vedic Description of the End of Time as a Great Fire Consuming Creation

Hindu scriptures describe the end of cosmic time as a phase of intense fire, where the universe is burned away before dissolution.

1. In the Vishnu Purana, it is said that:

- At the end of a cosmic cycle, seven suns will appear in the sky, scorching the Earth.
- The seas will boil, mountains will melt, and the entire cosmos will be engulfed in flames.
- This fire will consume all material existence before the universe collapses back into its primordial state.

2. In the Bhagavata Purana, it is described that:

- Lord Shiva, in his destructive aspect, will unleash cosmic flames, burning everything to ashes.
- These flames will rise from the core of the universe, spreading through the three worlds, dissolving all creation.

3. The Rig Veda speaks of an ultimate fire, Agni, that is responsible for cosmic dissolution:

- This fire is not simply destruction but also purification, resetting the universe for the next cycle.

This concept bears a striking resemblance to modern theories of energy release at the end of the universe, particularly the evaporation of black holes and high-energy phase transitions in physics.

Does This Align with the Idea of Hawking Radiation Evaporating Black Holes?

One of the most fascinating parallels between the Vedic firestorm and modern physics is Hawking radiation—the theoretical process by which black holes slowly lose energy and eventually evaporate.

1. The Fire of Black Hole Evaporation

- Stephen Hawking proposed that black holes are not truly eternal—they lose mass over time by emitting radiation.
- This process is extremely slow for large black holes, but near the end of their life, it accelerates dramatically.
- As a black hole shrinks, it emits increasing amounts of energy, leading to a final explosion—a burst of fire-like radiation before it vanishes completely.

This process closely resembles the Vedic idea that, before ultimate dissolution, a great fire will consume creation.

2. The Universe as a Giant Black Hole?

- Some physicists speculate that if the entire universe behaves like a massive black hole, it too may undergo a final evaporation phase.
- If true, this would mean that the universe will not end in a cold, dark state (Heat Death) but in a final burst of energy, just as the Vedas describe.

Could it be that Hindu cosmology's firestorm is a metaphor for the last burst of energy from evaporating cosmic structures, including black holes and perhaps even the universe itself?

Could the Universe End in a Final Burst of Energy?

Several modern scientific theories suggest that the universe might end not in stillness, but in a high-energy event, much like the Vedic descriptions of fire.

1. The Big Rip: A Universe Torn Apart by Energy

- Some scientists propose that dark energy will continue accelerating the expansion of the universe.
- Eventually, this energy could grow so strong that it overcomes all forces holding matter together.
- In this case, the universe would not fade away gradually, but explode in a final, high-energy event.

This theory aligns with the Vedic concept of the cosmic fire consuming everything before Pralaya.

2. Vacuum Decay: A Sudden, Universe-Wide Firestorm

- In quantum physics, some theories suggest that our universe exists in a "false vacuum"—a temporary stable state.
- If a quantum fluctuation were to create a more stable energy state, a bubble of "true vacuum" could form.
- This bubble would expand at the speed of light, annihilating everything in an instant—a true cosmic firestorm.

This idea matches the Hindu concept of a sudden, all-consuming fire at the end of time.

Conclusion: Is Cosmic Fire the Final Stage of Dissolution?

The idea that the universe ends in fire is not just a religious metaphor—it aligns with real astrophysical possibilities.

- The Vedic firestorm parallels the final burst of Hawking

radiation from black hole evaporation.

- The Big Rip scenario suggests that energy could one day consume the universe in a violent event.
- Vacuum decay presents the possibility of an all-consuming cosmic transformation, matching ancient descriptions of fiery dissolution.

If modern science is correct, then the ancients may have intuitively understood that fire—representing energy transformation—is the final step before complete dissolution.

In the next section, we will explore what happens after Pralaya: Does the universe truly end, or does it reset and start a new cycle? We will examine the cyclic nature of the cosmos and whether science supports the idea of infinite rebirth.

4.7 What Lies Beyond Pralaya? The Cyclic Universe and Eternal Renewal

If Maha Pralaya is the ultimate dissolution of everything—where time, space, and matter collapse into nothingness—does this mean existence comes to a permanent end? Or does it mark the beginning of a new cycle, a fresh cosmic dawn?

Hindu cosmology firmly holds that Pralaya is not the final conclusion, but rather a transition—a phase between the end of one cosmic cycle and the birth of another. The universe is believed to move through an infinite loop of creation, existence, dissolution, and rebirth.

Modern physics is now exploring similar ideas, questioning whether the Big Bang was truly the first moment of creation or if it was simply the latest in a long sequence of cosmic cycles. The oscillating universe model, the Big Bounce theory, and multiverse hypotheses all suggest that the cosmos may be far older and more cyclical than previously thought.

Could it be that the ancient sages, who envisioned endless cycles of dissolution and renewal, were intuitively describing the very nature of reality—one that modern science is only now beginning to understand?

Does Maha Pralaya Mark the End, or Does It Lead to a Rebirth?

Hindu scriptures describe Maha Pralaya as the return of everything—physical and non-physical—into a formless, unmanifest state (Mula Prakriti). This is not an absolute void but a cosmic potential, a seed of future creation.

1. In the Bhagavata Purana, it is said that:

- When the universe dissolves, only Narayana (the supreme consciousness) remains, resting in the causal ocean.
- From this state of infinite stillness, a new universe is born when the time is right.

2. The Vishnu Purana describes that:

- After Maha Pralaya, the Supreme Being breathes out once more, creating a fresh cosmos, just as the previous one was absorbed back into the infinite.
- This imagery aligns with the expansion and contraction models of the universe described in physics.

3. This cyclical concept is central to Hindu philosophy:

- Everything that exists must one day dissolve, but dissolution is only the prelude to a new beginning.
- Maha Pralaya is not the final end, but the necessary pause before cosmic renewal.

This vision of eternal recurrence finds remarkable parallels in modern cosmology.

The Oscillating Universe Model: Does Scientific Evidence Support an Eternal Cycle?

Scientists have long debated whether the universe had a single beginning or if it has been expanding and contracting in an eternal cycle.

1. The Big Bounce Hypothesis: A Scientific Parallel to Cosmic Renewal

- According to this theory, the universe may not end in a singularity but instead collapse and bounce back into another expansion phase.
- This would mean that each Big Bang is simply the latest in a never-ending sequence of universal cycles.
- If true, this is almost identical to the Hindu concept of Kalpas—cosmic cycles of birth, existence, and dissolution.

2. The Conformal Cyclic Cosmology Model (CCC)

- Proposed by Roger Penrose, CCC suggests that our universe is just one of many past universes, and future universes will emerge from its dissolution.
- He argues that the end of one universe merges seamlessly into the beginning of another.
- This theory strongly echoes the Vedic notion that the universe breathes in and out, moving from creation to dissolution and back again.

3. The Multiverse and Eternal Inflation

- Some theories suggest that our universe is just one in an endless "foam" of universes.

- If true, then new universes are constantly forming even as old ones dissolve, similar to how Hindu cosmology describes multiple Brahmas overseeing different cosmic cycles.
- This means that Pralaya may apply not just to our universe, but to infinite others.

These scientific perspectives lend weight to the idea that the cosmos is not a one-time event but an ongoing, self-renewing system—just as the ancient texts describe.

If Pralaya is Real, Are We Destined to Witness the Next Cosmic Rebirth?

The notion that our universe could dissolve and be reborn leads to an intriguing question: If cycles of cosmic creation and dissolution are real, could we ever witness them?

1. Timescales Beyond Human Comprehension

- The vast scales of time involved—millions, billions, and even trillions of years—mean that no single being could survive to see an entire cosmic cycle.
- However, if reincarnation or consciousness exists beyond the physical realm, as Hindu philosophy suggests, then could some essence of awareness persist through multiple cycles?

2. The Universe's Rebirth May Already Be Underway

- If the Big Bounce or cyclic universe models are correct, the seeds of the next cosmic rebirth may already be forming.
- Just as galaxies constantly reshape, dissolve, and reform, the larger universe may be undergoing a slow transformation toward its next phase.

3. Our Existence May Be a Memory of Past Universes

- Some physicists theorize that traces of past cosmic cycles could be embedded in the fabric of space-time.
- If so, then the universe we inhabit today may not be the first—it may simply be the latest iteration in an endless cosmic story.

These ideas suggest that Pralaya is not a singular event but a principle of nature—one that governs the infinite dance of existence.

Conclusion: The Eternal Cycle of the Universe

Whether through the lens of ancient wisdom or modern physics, the conclusion remains the same: existence is cyclic, and nothing truly ends.

- Maha Pralaya does not mark the absolute end, but the return to a state of unmanifest potential.
- The oscillating universe models, Big Bounce theories, and multiverse ideas align with the Hindu concept of eternal cosmic cycles.
- If Pralaya is real, then every universe that ends will inevitably give rise to a new one, continuing the infinite cycle of birth, dissolution, and rebirth.

If this is true, then the universe was never truly "created" in a singular event, nor will it ever be "destroyed" permanently—it simply moves through its endless cosmic rhythm, just as the ancients envisioned.

5: Time's End – Kaal Bhairav and the Distortion of Space-Time

In the previous chapter, we explored the dissolution of the universe through Pralaya, questioning whether cosmic destruction is the ultimate end or merely a transition into a new phase of existence. But if the universe undergoes cycles of creation and dissolution, what happens to time itself?

Time is perhaps the most mysterious aspect of reality—it governs existence, dictates change, and yet remains elusive in its true nature. Ancient Hindu cosmology recognized time not as a mere measurement but as a force—a relentless, consuming entity that dictates the life and death of all things. In its fiercest form, time is personified as Kaal Bhairav, the fierce manifestation of Shiva, the devourer of time itself.

In modern physics, black holes provide a startling parallel to Kaal Bhairav's legend. Near these extreme gravitational wells, time slows, stretches, and eventually seems to freeze at the event horizon—a point beyond which time ceases to function as we understand it. Inside a singularity, our current laws of physics collapse, and time itself may cease to exist.

Could it be that the ancients, in their mythological depictions, were intuitively describing the same phenomenon that modern physicists are only now uncovering? Does time truly end in black holes, or does it take on a new form beyond our comprehension? In this chapter, we explore the parallels between Kaal Bhairav, time's destruction, and the mysteries of time dilation, gravitational collapse, and the very nature of existence itself.

5.1 Kaal Bhairav: The Destroyer of Time

Throughout human history, different cultures have personified time as an unstoppable force—one that creates, preserves, and ultimately consumes everything. In Hindu cosmology, this force is represented by Kaal Bhairav, a fierce and enigmatic deity who is not just a guardian of time but its destroyer.

The very name Kaal Bhairav holds deep significance:

- "Kaal" means time or death, emphasizing his dominion over the temporal cycle.
- "Bhairav" means terrifying or beyond fear, indicating that he exists outside the limitations of time and decay.

Unlike other deities who symbolize creation or preservation, Kaal Bhairav is the aspect of Shiva that dismantles the illusion of permanence. His presence reminds us that all things—planets, galaxies, and even the universe itself—are temporary manifestations within the flow of time. When the time comes, everything must dissolve back into the great cosmic void, similar to the concept of Pralaya.

Why is Kaal Bhairav Seen as the Lord of Time's End?

Hindu mythology portrays Kaal Bhairav as the ultimate force that consumes all things, bringing an end to cycles of existence. He is often depicted with a garland of skulls, symbolizing the countless lives and ages that have already perished under the force of time.

According to ancient scriptures:

- He is beyond past, present, and future, existing outside the boundaries of time itself.
- He is invoked to dissolve illusions, breaking the attachments that bind beings to the cycle of birth and death.
- He is both terrifying and liberating, for while he destroys, he also frees existence from suffering and limitation.

This vision of time as a force that not only moves forward but also collapses upon itself is strikingly similar to modern physics' understanding of time near black holes. In the extreme gravity of a singularity, time does not function linearly—it slows, stretches, and ultimately ceases to exist as we perceive it.

Is Kaal Bhairav merely a mythological representation of black hole physics and the end of time as we know it?

How Kaal Bhairav's Role Represents the Relentless Force of Time

In many Hindu traditions, Kaal Bhairav is not just an abstract deity but an embodiment of the cosmic principle of time's dissolution. He is associated with:

1. The Destruction of Illusion (Maya)

- Time creates the illusion of permanence, but Kaal Bhairav reminds us that all things must decay and transform.
- This aligns with the Second Law of Thermodynamics, which states that entropy (disorder) must always increase, leading to the inevitable breakdown of all structures.

2. The Collapse of Reality into the Void

- In some interpretations, Kaal Bhairav is seen as the force that reduces all physical existence to pure energy or nothingness.
- This parallels the fate of matter inside a black hole, where it collapses into an infinitely dense singularity, effectively erasing its physical form.

3. The Guardian of Time's Threshold

- Temples dedicated to Kaal Bhairav are often located at boundaries—between cities, realms, and even dimensions.
- This reflects how black holes mark the threshold between

known physics and the unknown—where time and space break down.

Thus, Kaal Bhairav is not merely a deity of destruction but a symbolic representation of time's final act—the moment when reality itself collapses, just as the universe may one day dissolve into the fabric of cosmic emptiness.

The Connection Between Time's Destruction, Pralaya, and the Fate of the Universe

Just as Kaal Bhairav destroys all within time's grasp, the concept of Pralaya describes the dissolution of the cosmos itself. The two are intimately linked:

1. Maha Pralaya and the Ultimate End of Time

- Hindu cosmology states that at the end of each cosmic cycle, time itself dissolves into the unmanifest (Mula Prakriti).
- This reflects what physicists predict may happen if the universe undergoes a Big Crunch—where time, space, and matter collapse into a singularity.

2. Black Holes as Cosmic Timekeepers

- Just as Kaal Bhairav devours time, black holes trap light and matter, stretching time infinitely near their event horizons.
- Once something crosses the event horizon, time slows to a halt for an external observer—just as it does when Kaal Bhairav consumes existence.

3. The Ultimate Fate of the Universe

- If the universe continues expanding indefinitely, time as we know it may cease to have meaning in a cold, motionless cosmos (Heat Death).

- If the universe collapses into itself (Big Crunch), time itself could reverse or stop altogether, mirroring the idea that Kaal Bhairav dissolves time into its source.

Through this lens, Kaal Bhairav is not just a mythical figure but an early representation of deep cosmic truths—concepts that modern physics is still unraveling today.

Conclusion: Kaal Bhairav as the Final Gatekeeper of Time

Kaal Bhairav represents the most fundamental truth of existence—that time, no matter how powerful, is not infinite.

- He embodies time's relentless forward march, yet also its ultimate end.
- His legend parallels the scientific understanding of time distortion near black holes and cosmic singularities.
- He is the force that erases reality as we know it, much like gravitational collapse might one day erase the very fabric of space-time itself.

If Kaal Bhairav is indeed the destroyer of time, does he also hold the secret to what lies beyond it?

In the next section, we will explore how modern physics describes time near black holes—why time slows, stretches, and even halts at singularities, and how these ideas mirror ancient descriptions of time's dissolution.

5.2 Time and Gravity: How Black Holes Distort Reality

In our daily experience, time seems to flow at a constant rate—every second ticks forward uniformly, unchanging and absolute. However, modern physics has shattered this notion, revealing that time is not an independent, fixed entity but something that bends, stretches, and slows under extreme gravitational forces.

This concept, first introduced by Albert Einstein's theory of General Relativity, finds its most dramatic expression near black holes—cosmic objects so dense that they distort both space and time in ways that defy human intuition.

As we explore black holes and their effect on time, we find striking parallels with Hindu cosmology, where time is not a linear sequence but a flexible and dissolvable force, embodied in the legend of Kaal Bhairav. If black holes can slow time, halt it, or even make it vanish beyond a certain threshold, then could they be the very manifestations of time's dissolution—mirroring Kaal Bhairav's cosmic role?

The Nature of Time as Relative, Not Absolute – A Concept Modern Physics Confirms

For much of human history, time was considered absolute and unchanging—a universal clock ticking uniformly for all observers. This idea was deeply rooted in classical physics, until Einstein's theories of Special and General Relativity revealed that:

- **Time is relative**—it flows differently depending on an observer's speed and gravitational environment.
- **Stronger gravity slows down time**—meaning that time moves slower for objects near massive gravitational fields.
- **There is no single, universal "now"**—each observer experiences time differently based on their motion and location in space.

This means that time is not a fixed background against which events unfold—it is a dynamic entity that changes depending on where and how we measure it.

These principles find their most extreme expression in black holes, where gravity is so intense that time itself behaves in bizarre ways, appearing to slow down, stretch infinitely, or even stop altogether.

How Intense Gravity Near Black Holes Slows Down Time and Stretches Space

1. Time Dilation Near a Black Hole

- As an object approaches a black hole, an external observer would see its clock running slower and slower.
- The closer the object gets to the black hole's event horizon, the more time dilates—stretched out by the immense gravitational pull.
- If someone were to fall into a black hole, they would experience time normally, but to an external observer, they would appear to slow down, moving increasingly slower until they freeze at the event horizon.

This matches the legend of Kaal Bhairav, who is said to trap beings in timelessness, existing at the threshold where time loses meaning.

2. Space-Time Warping: The "Rubber Sheet" Effect

- Imagine space-time as a rubber sheet—when a massive object like a black hole sits on it, it creates a deep well in space-time.
- The closer something gets to this well, the more stretched and distorted its passage through time becomes.
- This is why, from an external perspective, time appears to slow to a halt near a black hole.

These effects are not just theoretical—they have been observed in real astronomical phenomena, confirming that time is a malleable, dynamic force that can be manipulated by gravity.

The Event Horizon: A Boundary Beyond Which Time Effectively Ceases to Exist

1. The Point of No Return

- The event horizon of a black hole is the boundary beyond which nothing—not even light—can escape.
- Anything crossing this threshold is effectively removed from the observable universe, as information from inside can never reach an outside observer.

2. Time Freezes at the Event Horizon

- For an observer watching from a safe distance, an object falling into a black hole appears to slow down indefinitely.
- At the event horizon, time stretches infinitely, meaning that the falling object appears frozen in time, never truly vanishing.
- But for the object itself, time proceeds normally—it simply crosses into a region where time has lost meaning for the outside world.

This closely resembles the philosophical and mythological descriptions of Kaal Bhairav, who stands at the threshold of time and eternity.

3. Does Time End Inside a Black Hole?

- Inside the event horizon, all paths in space-time lead inward toward the singularity, where our current understanding of physics breaks down.
- Many theories suggest that time itself ceases to exist within a singularity, as space-time collapses into an infinitely dense point.

This is eerily similar to the concept of Pralaya, where time dissolves into a formless, unmanifest state before the next cosmic cycle begins.

Conclusion: Black Holes as the Cosmic Agents of Time's End

Black holes present a modern scientific framework for understanding the distortion and potential dissolution of time, mirroring the Hindu notion of Kaal Bhairav as the devourer of time.

- Einstein's theory confirms that time slows down near massive gravitational fields, aligning with ancient descriptions of time's malleability.
- Black holes stretch and distort time, appearing to trap objects in an endless moment, just as Kaal Bhairav represents the threshold where time dissolves.
- Inside a black hole, time may cease to exist entirely, paralleling the cosmic dissolution described in Hindu philosophy.

If time can stretch, slow down, or even stop in extreme conditions, does this mean time itself is an illusion? Or is it merely a property of the universe that transforms under extreme forces?

In the next section, we will explore how ancient insights into time loops, time dilation, and gravitational collapse anticipated the most advanced ideas of modern theoretical physics.

5.3 The Legend of Kaal Bhairav and Time Dilation Near Black Holes

The concept of time slowing, stopping, or being completely dissolved is not just a feature of modern physics—it has existed in Hindu cosmology for millennia. The legend of Kaal Bhairav, the terrifying manifestation of Shiva, presents a vision of time that is not linear, absolute, or even permanent, but something that can be manipulated, consumed, or transcended altogether.

In modern physics, black holes demonstrate these very properties. As Einstein's General Theory of Relativity shows, the closer one gets to an extreme gravitational field, the slower time flows—until, at a singularity, time ceases to exist as we understand it. Could it be that the ancient sages intuitively understood the concept of time dilation near massive objects, encoding this knowledge into mythological narratives?

Ancient Stories of Kaal Bhairav as a Force Beyond the Constraints of Time

Kaal Bhairav is one of the most enigmatic and feared deities in Hinduism, often depicted as a fierce, dark, and formless being who governs time, destruction, and the dissolution of reality. Unlike the conventional understanding of time as an endless linear flow, Kaal Bhairav exists outside of time, consuming and transcending it.

1. The Story of Kaal Bhairav and Time's End

One of the most well-known legends describes how Kaal Bhairav decapitated Brahma's fifth head, symbolizing the destruction of ego and time itself.

- After this event, Bhairav was cursed to wander timelessly, transcending the past, present, and future.
- This resonates deeply with the idea that, near a black hole's event horizon, time slows down drastically—almost freezing, creating an illusion of timelessness.
- If someone were to observe Kaal Bhairav, they might see him as an eternal presence, much like an object at a black hole's horizon that appears frozen in time from an external perspective.

2. Kaal Bhairav as the Guardian of Time's Threshold

Kaal Bhairav is often worshipped at city gates, temple entrances, and cosmic boundaries, symbolizing his role as the gatekeeper between different dimensions of time and existence.

- In physics, a black hole acts as a similar cosmic gateway—a boundary between the observable universe and the unknown.

- Just as Kaal Bhairav controls time's dissolution, a black hole governs the fate of time and space, determining what is lost forever beyond its event horizon.

Could these descriptions be early philosophical interpretations of gravitational time dilation and the distortion of space-time near singularities?

Do the Mythological Descriptions of Timelessness Align with Time Dilation Near Black Holes?

Time dilation, predicted by Einstein's General Theory of Relativity, describes how time slows in strong gravitational fields. The closer an object moves toward a black hole, the slower its clock ticks relative to an outside observer.

This effect bears striking similarities to the descriptions of Kaal Bhairav's existence:

1. The "Frozen Time" Illusion at the Event Horizon

- According to Hindu cosmology, Kaal Bhairav does not age, change, or decay—he exists in a state beyond time.
- In modern physics, a person falling toward a black hole would appear frozen at the event horizon, their time stretching infinitely from the perspective of an external observer.
- This means that both mythological and scientific descriptions of time near a cosmic threshold share a common theme: time slows, distorts, or ceases altogether.

2. Near-Light Speed and the Illusion of Immortality

- Ancient texts describe celestial beings who experience time differently from mortals.
- Modern physics confirms that near the speed of light, time

dilates—meaning that someone traveling at relativistic speeds would age much slower than those in a normal gravitational field.
- This suggests that the ancient vision of divine beings outside normal time constraints could be linked to extreme relativistic effects.

Could it be that Kaal Bhairav's legend encodes knowledge of time dilation—an understanding that powerful gravitational forces can stretch or halt time?

How Einstein's Theory of General Relativity Predicts That Time Stops at Singularities

Einstein's General Theory of Relativity provides a framework for understanding how time behaves near massive objects. Black holes, the densest known objects, create such extreme distortions in space-time that time slows down dramatically near them.

1. Approaching the Event Horizon:

- As an object moves closer to a black hole, an outside observer sees its time slowing down exponentially.
- The closer it gets to the event horizon, the slower its clock ticks relative to the rest of the universe.
- At the event horizon, time stretches infinitely, appearing to freeze in place.

2. Inside the Event Horizon:

- Once past the event horizon, all paths in space-time lead inward to the singularity.
- Time, as we understand it, loses meaning inside a singularity—it either ceases altogether or transitions into an unknown state.

This description is eerily similar to the idea of Kaal Bhairav consuming time itself. If black holes represent the physical manifestation of time's end, then Kaal Bhairav can be seen as a conceptual embodiment of this principle, a mythological personification of relativistic time effects.

Conclusion: Kaal Bhairav as the Mythological Representation of Time Dilation

The story of Kaal Bhairav aligns with the way black holes alter time, providing an ancient conceptual model for a phenomenon that modern physics is only now fully understanding.

- Kaal Bhairav's existence beyond time reflects the way time slows and effectively halts near a singularity.
- The event horizon's "frozen time" effect mirrors the mythological notion of Kaal Bhairav standing outside the normal flow of time.
- Black holes serve as cosmic gateways where time's laws break down, much like Kaal Bhairav is the guardian of time's dissolution.

If time can be manipulated, distorted, or even stopped under extreme conditions, then perhaps time itself is not an absolute entity but a property of the universe that can be reshaped.

In the next section, we will explore time loops, eternal recurrence, and the possibility that the universe itself experiences repeating cycles of time—an idea shared by both Hindu cosmology and modern physics.

5.4 Time Loops, Eternal Recurrence, and the Myth of Cyclic Time

The nature of time is one of the greatest mysteries in both ancient philosophy and modern physics. While many cultures perceive time as a straight path moving from the past to the future, Hindu cosmology describes time as inherently cyclic—a repeating sequence of creation, existence, destruction, and rebirth.

This idea of cyclic time aligns with modern cosmological theories, such as the oscillating universe hypothesis, the concept of eternal recurrence, and even the possibility of closed time-like curves (CTCs), which suggest that time might loop back onto itself. Could it be that ancient Indian sages, through their understanding of Kaal (time), were describing the same time loops and cyclic universe models that physicists are only now beginning to explore?

Furthermore, black holes, with their extreme warping of space-time, may hold the key to understanding whether time can be bent, looped, or reset—mirroring the very essence of Kaal Bhairav's legend as the force that controls and resets time.

The Hindu Perspective of Time as Cyclic, Not Linear – Matching the Oscillating Universe Theory

In Hindu cosmology, time is not an arrow moving in one direction, but a wheel—an endless cycle of cosmic events. This cycle is described through the Yugas, Kalpas, and Maha-Pralayas, indicating that time is a repeating sequence of creation and destruction, rather than a one-time event.

1. The Cycles of Yugas and Kalpas

- Hindu philosophy states that the universe passes through repeated cosmic ages (Yugas), eventually leading to a Pralaya (dissolution) before a new cycle begins.
- The concept of Kalpas (cosmic days of Brahma) describes a universe that is born, sustained, and then dissolved—only to be reborn in a new cosmic cycle.

2. The Oscillating Universe Theory

- Modern physics proposes that the universe may not have

begun with a single Big Bang, but instead undergoes an endless cycle of expansion and contraction.
- The Big Crunch, Big Bounce, and cyclic universe models suggest that the universe might collapse and then "reinflate," just like the Hindu concept of repeated creations and dissolutions.

This presents an intriguing parallel—both Hindu cosmology and modern science suggest that the past, present, and future may not be unique but part of an endless cycle of recurrence.
Could this mean that time itself loops rather than progresses linearly?

Scientific Discussions of Closed Time-Like Curves (CTCs) – Can Time Loop Onto Itself?

Modern physics has explored the possibility that time may not only be cyclical on a cosmic scale but might also form actual loops in space-time itself. These loops, known as Closed Time-Like Curves (CTCs), suggest that time can fold back onto itself, allowing for possible time loops or even time travel.

1. Theoretical Existence of Time Loops

- In Einstein's General Relativity, space-time is not fixed—it can be warped by gravity.
- Certain solutions to Einstein's equations allow for regions where time curves back onto itself, creating a loop.
- This means that in extreme conditions, an event could lead back to its own past—forming a cycle rather than a straight path.

2. Black Holes and Wormholes as Pathways for Time Loops

- Rotating black holes (Kerr black holes) theoretically allow for stable closed time-like curves.

- This means that, under the right conditions, a traveler inside such a black hole might experience time folding back, essentially looping through history.
- Wormholes—hypothetical tunnels in space-time—have also been proposed as potential structures where time loops could exist.

These concepts are eerily similar to the Hindu idea that time itself is not absolute but flexible—capable of repeating cycles and loops rather than a strict beginning and end.
Could black holes serve as natural manifestations of this time-looping property—acting as cosmic resets, just as Kaal Bhairav is said to reset time?

Are Black Holes Gateways to Repeating Time Cycles, Mirroring Kaal Bhairav's Legend?

1. Kaal Bhairav as the Guardian of Time's Reset

- In Hindu mythology, Kaal Bhairav is said to erase time, disrupt its continuity, and govern its flow outside human comprehension.
- This aligns with how black holes distort and manipulate time, effectively halting or resetting it from an external perspective.

2. Black Holes as Cosmic "Time Recycling Machines"

- If the universe follows a cyclic model, black holes may play a role in recycling cosmic matter and energy, preparing for the next cycle.
- Some physicists suggest that information is not lost in black holes but instead encoded and possibly released in another form—potentially in a future universe.

3. Do Black Holes Contain Traces of Past Universes?

- Some modern quantum theories suggest that remnants of past universes may survive in black hole singularities, carrying imprints of previous cycles.
- This would align with the idea that Kaal Bhairav preserves cosmic memory beyond ordinary time, ensuring that each cycle of existence is connected to the last.

If black holes are indeed gateways where time folds upon itself, allowing for cosmic resets, then Kaal Bhairav's symbolic role as the force that transcends and resets time may be a deeply intuitive representation of what physics is only beginning to theorize.

Conclusion: Is Time Truly a Loop?

The ancient Hindu view of time as cyclic aligns with cutting-edge theories in cosmology and physics, suggesting that time is not a straight arrow but part of an eternal recurrence.

- The Yuga cycles and Kalpas describe cosmic time as something that repeats infinitely, much like the oscillating universe theory suggests.
- Closed Time-Like Curves (CTCs) propose the possibility of time loops, meaning time might fold onto itself rather than moving in one direction.
- Black holes, with their extreme warping of space-time, may serve as natural pathways to these time loops, mirroring Kaal Bhairav's role as the force governing time's dissolution and reset.

If time is not linear, but a vast, repeating cycle, does this mean that everything that has happened will happen again? Does the destruction of one universe immediately lead to the birth of another?

In the next section, we will explore the ultimate boundary of time—the gravitational singularity. If time loops and stretches near black holes, what happens when time completely ceases? Is there something beyond time itself?

5.5 Gravitational Collapse: When Time Stops

Throughout this chapter, we have explored how time bends, slows, and loops under extreme gravitational conditions, particularly near black holes. But what happens when gravity becomes so intense that even time itself ceases to function? Does time reach an absolute stop, or does it transition into an entirely new state—one beyond human comprehension?

Hindu cosmology speaks of Pralaya, the dissolution of the universe, where time itself is absorbed into the unmanifest (Mula Prakriti), ceasing to exist in its conventional form. Modern physics, too, proposes a strikingly similar scenario: gravitational collapse. When a black hole forms, the fabric of space-time itself is dragged into its singularity, leading to a complete breakdown of the known laws of physics.

If Pralaya is the reset mechanism of the universe, could it also mean the dissolution of time itself? Could the end of time near a singularity be the same as the cosmic reset described in ancient texts?

What Happens Inside a Black Hole? Does Time Come to an Absolute Halt?

To answer this, we must understand the nature of a black hole's singularity—the point of infinite density where gravity is so strong that nothing, not even light, can escape.

1. The Event Horizon: The Edge of Time's Slowdown

- As an object approaches the event horizon of a black hole, an external observer sees its clock ticking slower and slower.
- At the event horizon, time appears to freeze for an external observer, meaning that from their perspective, the falling object never actually crosses the boundary.

2. Inside the Event Horizon: The Death of Time

- Once past the event horizon, all paths in space-time lead inward to the singularity.
- Here, gravity becomes so extreme that time ceases to function in any recognizable way.
- In Einstein's equations, time becomes just another spatial coordinate, meaning that "past" and "future" lose their meaning.

3. The Singularity: A Point Beyond Time and Space

- At the very center, all known physics breaks down.
- Time, as we measure it, is no longer meaningful—it doesn't exist in the way we understand it.
- Some theories suggest that the singularity is a "doorway" to a new kind of space-time or even a new universe.

This description is eerily similar to the Hindu idea of Maha Pralaya, where the universe dissolves into a formless, timeless state before rebirthing into a new cosmic cycle.
Could black holes be nature's version of Pralaya—zones where space and time cease to exist in their normal form, only to emerge in an unknown reality?

The Possibility That Time and Space Merge into a New State Beyond Human Understanding

When we think about the collapse of time inside a black hole, we are forced to consider the possibility that:

1. Time is not fundamental but emerges from deeper cosmic laws.

- Just as Pralaya dissolves all material existence into its root energy, a singularity may dissolve time into its fundamental essence—whatever that may be.

2. Black Holes as Cosmic Bridges

- Some physicists believe that black holes could serve as gateways to different universes, or even different states of existence.
- If so, then the singularity inside a black hole could represent a point where time resets and transforms into something new.

3. Does Time Emerge Again Elsewhere?

- If time dissolves inside a black hole, does it reappear elsewhere?
- Some speculative models, like the Big Bounce theory, suggest that a singularity could lead to a new Big Bang, echoing the idea that Pralaya is a temporary phase before creation begins again.

This concept strongly aligns with Hindu cosmology, where time does not "end" permanently but undergoes a transformation, returning when the next cycle begins.

If Pralaya Resets the Universe, Does Time Itself Dissolve or Transform into Something Else?

In Hindu thought, time is absorbed back into its source during Maha Pralaya, leaving only the unmanifest, timeless state before a new cycle begins. This raises the question:

- Does time truly disappear, or does it take a new form beyond our comprehension?
- Does time exist only when there is an observer, much like quantum mechanics suggests?
- If time stops inside a singularity, could it mean that the next universe's time begins anew?

Some modern theories propose that time itself may be an emergent property—not fundamental to reality, but a feature of the universe that arises under certain conditions.

- If this is true, then time, like matter and energy, is simply another aspect of creation that appears and disappears in cosmic cycles.
- Much like Pralaya suggests that all of creation dissolves and is reborn, perhaps time itself is also part of this rebirth.

Conclusion: The Singularity as the Ultimate Dissolution of Time

At its core, gravitational collapse into a singularity mirrors the concept of Pralaya—the dissolution of time and space itself.

- Inside a black hole, time ceases to exist as a meaningful concept, just as Pralaya absorbs time into its source.
- If the universe undergoes cycles of birth and destruction, could time itself also be cyclical, resetting with each cosmic rebirth?
- Perhaps, time does not "end" but transforms—becoming something else entirely, just as matter and energy transform under extreme conditions.

If black holes are singularities where time dissolves, could they be cosmic engines of rebirth, just as Pralaya is the reset mechanism of creation?

In the next section, we will explore the ultimate question: Is time truly an illusion? Or is it something deeper, waiting to be understood beyond the limits of human perception?

5.6 Does Time Truly End, or Does It Evolve?

Throughout this chapter, we have explored how time bends, stretches, loops, and ultimately dissolves in the presence of extreme gravitational forces. Black holes, with their immense warping of space-time, appear to hold the key to the ultimate fate of time itself. But the question remains: Does time truly end inside a singularity, or does it evolve into something beyond our understanding?

Hindu cosmology, particularly in the concept of Maha Pralaya, suggests that time is not permanently destroyed but rather reset—absorbed into a timeless state before re-emerging in a new cosmic cycle. Similarly, modern theoretical physics is beginning to propose that time may not simply vanish at a singularity but could transition into a new phase, possibly giving rise to another universe.

If this is true, then time is not an absolute entity with a beginning and end—it is a fluid, evolving force that changes forms, much like energy and matter. Could it be that Kaal Bhairav, as the guardian of time's destruction, also holds the secret to its rebirth?

If Time Stops at a Singularity, Is It Truly an End, or Does It Transition Into a Different State?

When an object falls into a black hole, it crosses the event horizon and is inevitably pulled toward the singularity. According to Einstein's General Relativity:

- Time stretches infinitely as it nears the singularity.
- From an external observer's point of view, the object never actually reaches the singularity—it appears frozen in time at the event horizon.
- Inside the event horizon, time and space swap roles—time no longer moves forward in a conventional sense.

This raises a profound question: Does time truly stop inside a singularity, or does it transition into something else?

1. The Evolution of Time Beyond the Singularity

- If singularities represent a state where space and time lose meaning, could it be that time doesn't end but transforms into a new kind of existence?
- Some scientists speculate that inside a black hole, our known universe ceases to function, but another reality—a new dimension of time and space—may emerge.
- This aligns with Hindu thought, where time dissolves in Pralaya, only to be reborn in the next cosmic cycle.

2. Is a Black Hole's Singularity the Birthplace of a New Universe?

- Theoretical physicists propose that the singularity inside a black hole may be more than just a point of infinite density—it may be a seed for a new universe.
- Some models suggest that when a black hole collapses, the singularity doesn't just vanish—it could explode outward on the "other side," creating a new expanding universe.
- This resembles the Big Bounce hypothesis, where each universe's collapse seeds the next one—mirroring the Hindu vision of Kalpas and cosmic rebirth.

If time can evolve beyond a singularity, then it is not truly destroyed—it simply enters a new phase, much like the cyclical nature of existence described in Hindu cosmology.

Theoretical Physics Suggests That Time Could Emerge Differently in a New Universe

Many modern theories challenge the idea that time has a singular beginning and an absolute end. Instead, time may be part of a continuous, evolving structure that undergoes transformation rather than destruction.

1. The Big Bounce Hypothesis: Time Resets with Each Cosmic Cycle

- Instead of a one-time Big Bang, the Big Bounce model suggests that the universe undergoes repeated expansions and contractions.
- If this is true, time does not end at a singularity—it is simply reset, beginning anew in each cosmic cycle.
- This is identical to the Vedic idea of Maha Pralaya, where time dissolves only to re-emerge when creation begins again.

2. Quantum Gravity and the Possibility of Timeless States

- Some quantum models suggest that time may not be fundamental to the universe at all.
- Instead, time may be an emergent property of space-time that arises under certain conditions, meaning it could dissolve and then reform in different ways.
- This echoes Hindu thought, where time exists only within manifest creation, dissolving back into the unmanifest (Mula Prakriti) during Pralaya.

3. Does the Multiverse Hold the Answer?

- Some theories propose that our universe is just one in a vast multiverse, where different universes exist with different versions of time.
- If true, this means time does not truly end—it simply continues in another form, in another universe.

All of these ideas suggest that time is not a fixed, absolute entity—it is a fluid phenomenon that can dissolve, reset, or transform.

Does the Legend of Kaal Bhairav as the Keeper of Time's Destruction Hint at the Secret of Time's Rebirth?

In Hindu tradition, Kaal Bhairav is not just the destroyer of time—he is also its guardian.

- He is depicted as fierce, consuming, and beyond time itself, yet he is also the force that governs the renewal of cosmic cycles.
- This dual role suggests that time's dissolution is not permanent but a necessary part of its evolution.
- If time were linear, it would simply end—but in both Hindu thought and modern physics, time seems to fold, loop, or restart in some form.

Could this mean that black holes—often viewed as cosmic endpoints—are actually functioning as time's recycling centers, much like Kaal Bhairav's mythological role?
Black Holes as Cosmic "Kaal Bhairavs"

- Just as Kaal Bhairav dissolves time, black holes stretch, distort, and absorb it.
- Just as Kaal Bhairav stands at the threshold between creation and destruction, black holes seem to be the boundary between one state of existence and another.
- Just as Kaal Bhairav is a force of transformation rather than absolute annihilation, black holes may not destroy time but evolve it into a new form.

If Hindu philosophy views Kaal Bhairav as the force that governs the cyclical nature of time, then perhaps black holes—rather than being "the end"—are part of an ongoing cosmic cycle, where time resets and begins again.

Conclusion: Time Is Not Destroyed, It Evolves

As we merge ancient wisdom with modern science, a powerful realization emerges:

- Time does not have a fixed beginning or end—it is fluid, capable of evolving into new forms.

- Singularities and black holes do not destroy time permanently—they may transition it into something else, possibly seeding new universes.
- Hindu cosmology's vision of Pralaya and cosmic rebirth aligns with scientific theories that time and space may reset in endless cycles.
- Kaal Bhairav's legend suggests that time is never truly lost—it simply transforms beyond human perception, waiting to emerge once again.

If black holes hold the key to time's evolution, then perhaps our understanding of time itself is incomplete. What we see as "the end" may simply be another doorway—one leading to a reality beyond the limits of our current knowledge.

In the next chapter, we will explore the deeper philosophical and scientific implications of time's transformation—what it means for our understanding of reality, existence, and the future of the cosmos itself.

6: The Cosmic Egg and the Universe's Rebirth

In the previous chapter, we explored the dissolution of time at the threshold of black hole singularities, questioning whether time itself ends or simply evolves into a new state. But if time is not truly lost, could the universe itself follow a similar pattern—dissolving only to be reborn again?

Ancient Hindu cosmology presents a strikingly similar vision of cyclical creation and dissolution. The concept of Hiranyagarbha, the Golden Womb, describes a cosmic seed from which all existence emerges—a notion that eerily parallels the Big Bang theory, where the universe expands from an infinitely dense singularity.

Modern physics has begun to explore whether black holes—often viewed as cosmic endpoints—could actually serve as gateways to new universes. Some theories suggest that what appears as the collapse of space-time in one universe could be the birth of another. If so, the death of our universe might not be an end but a transition—just as the ancient sages described the cosmic cycle of creation, preservation, and dissolution.

Could black holes be the Hiranyagarbha of modern cosmology, containing within them the seeds of future universes? Is our existence part of an infinite sequence of cosmic rebirths, where time, space, and matter continuously collapse and re-emerge in an eternal cycle?

In this chapter, we explore how the ancient concept of Hiranyagarbha aligns with modern theories of the Big Bang, black hole cosmology, and the cyclical nature of the universe, seeking to uncover whether our reality is a singular event or part of a never-ending cosmic rhythm.

6.1 Hiranyagarbha: The Golden Womb and the Birth of the Cosmos

In Hindu cosmology, the origin of the universe is described through the concept of Hiranyagarbha, meaning "Golden Womb" or "Golden Egg." This primordial cosmic entity is seen as the seed of all creation, floating in the void before the universe took shape. It is described in the Rig Veda as the source of all existence, containing within it the potential for space, time, and matter—much like the singularity before the Big Bang in modern physics.

Could Hiranyagarbha be an ancient representation of the compressed state of energy and information that existed before the Big Bang? Was this golden egg not just a metaphor but a deeply intuitive description of the universe's birth, remarkably similar to what science now theorizes?

What is Hiranyagarbha in Hindu Cosmology?

Hiranyagarbha is one of the most profound cosmological concepts in the Vedic tradition, representing the original state of the universe before it expanded into existence. In the Rig Veda (10.121), it is described as:

"In the beginning was Hiranyagarbha, the Golden Womb, the one lord of all that was born. He established the Earth and the sky. Who is the deity we shall worship with offerings?"

This passage suggests that:

- Hiranyagarbha existed before time, space, and material creation.
- It contained all of existence in a potential, unmanifest form.
- It eventually "broke open," leading to the structured cosmos we observe today.

This description is strikingly similar to the modern idea that the universe once existed in an incredibly dense, high-energy state before expanding into the vast cosmos we now inhabit.

Hiranyagarbha as the Source of Space and Time

According to Hindu cosmology, Hiranyagarbha existed in a timeless, spaceless void before creation. This echoes modern physics' understanding that:

- Before the Big Bang, space and time as we know them did not exist.
- The universe was in an extremely dense state, with all of existence compressed into a single point of infinite potential.
- The emergence of space and time was not an explosion but an unfolding expansion from a singularity—just as Hiranyagarbha "opened" to create the cosmos.

Could it be that the Vedic sages, through deep contemplation, arrived at an understanding of cosmic origins long before modern science?

How the Golden Womb Represents the Primordial Singularity Before Cosmic Creation

The idea that all of creation was once contained within a single, unified state of potential is central to both Hindu cosmology and modern astrophysics.

1. Hiranyagarbha as the Compressed State of Existence

- Just as the Big Bang theory proposes that the universe was once concentrated into a singularity, the Vedic tradition describes Hiranyagarbha as a condensed source of all matter, energy, and consciousness.
- Before the universe took shape, everything existed in a primordial state—undivided, timeless, and infinitely dense.

2. The "Breaking Open" of Hiranyagarbha and the Big Bang

- According to Hindu mythology, Hiranyagarbha "broke open" and expanded to form the cosmos.
- This resembles the Big Bang, where the initial singularity expanded, giving rise to space, time, and the fundamental forces of nature.

3. Creation as an Expanding Process

- The universe today is still expanding, just as Hiranyagarbha continuously unfolds into the vast cosmos.
- This expansion does not mean an explosion but a highly ordered process, mirroring the Vedic idea that creation is a structured and rhythmic event.

These parallels suggest that Hiranyagarbha may not be just a symbolic representation of creation but an early intuitive explanation of how the universe came into existence—one that aligns with modern scientific understanding.

Does Hiranyagarbha Symbolize the Compressed State of Energy Before the Big Bang?

In modern physics, the Big Bang theory suggests that:

- The universe began as an incredibly dense and hot singularity.
- It contained all the matter and energy that would eventually expand and form galaxies, stars, and life.
- Before this moment, space and time as we understand them did not exist.

This description aligns perfectly with the concept of Hiranyagarbha in Hindu texts:

Concept	Hindu Cosmology (Hiranyagarbha)	Modern Science (Big Bang)
Primordial State	Unmanifest cosmic potential before creation	Singularity containing all energy before expansion
Contained Reality	All of existence within the Golden Womb	Universe compressed into a single point
Expansion Process	"Breaks open" to form the universe	Rapid expansion of space and time
Timeless Origin	Exists before time and space	Space and time emerge after singularity expands
Structured Evolution	Creation unfolds in cycles	Cosmic expansion follows physical laws

This remarkable convergence of ancient wisdom and modern physics suggests that Hiranyagarbha may be one of the earliest conceptualizations of the universe's birth. It portrays the idea that the cosmos emerges not from chaos but from a structured, ordered process—one that modern astrophysics is only now beginning to decode.

Conclusion: The Cosmic Egg as a Scientific Insight

The Hiranyagarbha concept presents a deeply insightful vision of cosmic creation, one that aligns with the Big Bang theory and the idea of a primordial singularity.

- Hiranyagarbha represents a compressed cosmic seed, much like the singularity before the Big Bang.
- The "breaking open" of the Golden Womb parallels the expansion of the universe into space and time.
- Both Hindu thought and modern physics describe a universe that emerges from a highly ordered initial state, rather than a

chaotic explosion.

If ancient Indian philosophy accurately described the birth of the cosmos thousands of years ago, could it also hold clues about what happens after the universe collapses? Is creation a singular event, or does the universe go through infinite cycles of rebirth?

In the next section, we will explore how modern physics, through black holes and cosmic portals, may provide evidence for a cyclical universe—one that has no true beginning or end, but instead moves through infinite cycles, just as the ancients described.

6.2 The Big Bang Theory and Its Connection to Ancient Cosmology

For centuries, human civilizations have pondered the origin of the universe—was it created in a single event, or has it always existed? Modern physics now suggests that the universe began with the Big Bang, a rapid expansion from an infinitely dense point of energy and matter. This theory, developed through observational evidence and mathematical models, describes how the cosmos unfolded from an incredibly hot and compact state into the vast, evolving universe we see today.

However, the idea of a cosmic seed giving birth to the universe is not unique to modern science. Hindu cosmology, through the concept of Hiranyagarbha—the Golden Womb, describes an uncannily similar vision of the universe's birth. Could it be that ancient Indian sages, through deep contemplation and observation, intuited the fundamental principles of cosmic creation long before the advent of scientific models?

The Scientific Model: How the Universe Expanded from an Infinitely Dense Point

The Big Bang Theory is the leading scientific explanation for the origin of the universe. It suggests that:

- The universe began as a singularity—a point of infinite density and temperature.
- Approximately 13.8 billion years ago, this singularity began expanding rapidly.
- This expansion led to the formation of space, time, and fundamental particles.
- The universe continues to expand today, as observed through the redshift of galaxies.

This model does not suggest that the Big Bang was an explosion in space but rather an expansion of space itself. There was no "before" the Big Bang in the conventional sense because time itself emerged from this event.

Strikingly, this description of a primordial state leading to cosmic expansion mirrors the Vedic idea of Hiranyagarbha—an infinite potential that "breaks open" to create the structured universe.

Parallels Between Hiranyagarbha's Expansion and the Inflationary Model of the Early Universe

One of the most significant aspects of modern cosmology is the inflationary model, which describes how the universe expanded exponentially within a fraction of a second after the Big Bang.

1. Inflationary Expansion in Science

- The universe expanded faster than the speed of light in its earliest moments, smoothing out irregularities.
- This rapid expansion transformed the initial singularity into the vast cosmic structure we see today.
- After inflation, the universe continued to expand at a slower rate, allowing matter and galaxies to form.

2. Hiranyagarbha's Expansion in Hindu Cosmology

- The Golden Womb is described as the source of all existence, containing everything in a latent, undivided state.
- Upon creation, Hiranyagarbha "breaks open," giving birth to space, time, and the physical universe.
- This description is remarkably similar to inflation, where the universe emerges rapidly from a unified state into a structured cosmos.

The Rig Veda describes Hiranyagarbha as:
"In the beginning was the Golden Womb, the source of all. It was the first lord of all beings, supporting Earth and sky. From it arose the cosmos."
This resonates with modern physics' description of the early universe, where space, time, and matter originated from a compressed, singular state.

Could the Ancients Have Intuitively Understood the Birth of the Universe as Described in Modern Physics?

The similarities between the Hindu concept of creation and the Big Bang model raise an intriguing question:

- Did the ancient Indian sages have knowledge of cosmic evolution that aligns with modern scientific discoveries?
- Was Hiranyagarbha a philosophical intuition of the expanding universe?

Although ancient texts do not describe physics in mathematical terms, they provide metaphysical insights that closely parallel modern theories. Some possibilities include:

1. Observation of Cosmic Patterns

- Ancient sages may have observed the cyclic nature of celestial bodies, planetary movements, and cosmic rhythms.
- This may have led them to propose a cosmic seed that expands

into the structured universe, much like a plant growing from a seed.

2. Symbolic Representation of Scientific Truths

- The Vedic description of the cosmos emerging from a single, undivided source could be an intuitive understanding of the universe emerging from a singularity.
- While modern science uses mathematics and observation, ancient thought used metaphor and symbolism to describe the same reality.

3. The Concept of Spontaneous Creation

- Some Hindu texts describe the universe emerging spontaneously from a timeless, eternal source, much like the vacuum fluctuations in quantum physics.
- The idea that space and time were born together from a primordial source aligns with modern physics' idea that time began with the Big Bang.

Conclusion: A Shared Vision of Cosmic Creation

The comparison between Hiranyagarbha and the Big Bang suggests that ancient Hindu cosmology may have anticipated key ideas of modern astrophysics.

- Both describe the universe emerging from an initial, undivided, high-energy state.
- Both suggest that creation is not random but follows a structured process.
- Both imply that time and space originated from this cosmic seed.

If these parallels exist, could the ancients have possessed knowledge that aligns with modern discoveries, encoded in spiritual and philosophical language?

In the next section, we will explore whether black holes—often seen as cosmic endpoints—could instead be portals to new universes, serving as modern equivalents of the cosmic womb described in ancient texts.

6.3 Black Holes as Cosmic Portals: Could They Give Birth to New Universes?

In our exploration of cosmic origins, we have seen how the Vedic concept of Hiranyagarbha parallels the modern Big Bang theory—a primordial seed giving birth to the universe. But if the universe had a beginning, does it also have an end? And could that end lead to the birth of a new universe?

Modern physics is beginning to entertain a radical hypothesis: black holes—once thought to be cosmic dead ends—may actually serve as gateways to new universes. The idea that each black hole singularity could seed a separate cosmos is gaining traction, raising profound questions about the nature of reality, space-time, and the potential for an infinite multiverse.

What is fascinating is that Hindu cosmology may have anticipated this idea. Some Vedic and Puranic texts describe the universe as being cyclically reborn through cosmic "wombs" or gateways—concepts that align with modern discussions on black holes as potential portals to new realms. Could black holes be modern scientific equivalents of Hiranyagarbha, the Golden Womb of creation?

The Hypothesis That Black Hole Singularities Could Seed New Universes

A black hole is formed when a massive star collapses under its own gravity, creating a singularity—a point of infinite density where space and time break down. Traditionally, singularities have been considered cosmic dead ends, but recent theories suggest they might instead be gateways to the birth of new universes.

1. Singularities as Cosmic Seeds

- Some physicists propose that the extreme density and energy inside a black hole singularity could give rise to a new expanding universe.
- If a singularity could "burst" outward in another dimension, it might lead to a Big Bang-like expansion, creating a new space-time bubble separate from our own.

2. Theoretical Models Supporting This Idea

- **Lee Smolin's "Cosmological Natural Selection" Hypothesis:** This theory suggests that universes evolve like living organisms, with each black hole spawning a new cosmos. Each "baby universe" inherits physical laws from its parent, allowing for the possibility of an infinite lineage of universes.
- **Roger Penrose's Cyclic Conformal Cosmology (CCC):** Suggests that the death of one universe through black hole evaporation may lead to the birth of another.
- **The "White Hole" Hypothesis:** Some physicists propose that what we see as a black hole from one perspective may be a white hole—a cosmic eruption—on the other side, forming a new universe.

These ideas suggest that black holes are not just destructive forces but could be the very wombs of cosmic rebirth, much like Hiranyagarbha is described as the golden egg of creation.

The Idea of Universe Reproduction: Each Black Hole Forming Its Own Separate Cosmos

If black holes do indeed seed new universes, this means that the multiverse is not just a theoretical construct—it may be a natural consequence of black hole formation.

1. Black Holes as "Cosmic Parents"

- Every universe could give birth to multiple offspring through black holes, each forming a distinct cosmos with its own laws of physics.
- This means that our own universe may have originated from the singularity of a black hole in a parent universe.

2. Self-Replicating Universes and the Multiverse Theory

- The idea that black holes create new universes suggests that existence is not a one-time event but an endless, self-sustaining cycle.
- This aligns with the Hindu idea of Yugas and Kalpas—cycles of creation and dissolution, endlessly repeating.

Could the ancient Indian sages have glimpsed this idea through their cosmological vision of a cyclically reborn universe?

Do Hindu Texts Hint at Black Holes as Cosmic Gateways, Mirroring the Role of Hiranyagarbha?

Hindu cosmology is rich with descriptions of cosmic cycles, vast realms, and portals between different planes of existence. Some of these descriptions bear an uncanny resemblance to modern theories of black holes as gateways to other universes.

1. Hiranyagarbha and the Cosmic Womb as a Black Hole Portal

- The Golden Womb (Hiranyagarbha) is described as a cosmic seed containing all of creation before it unfolds.
- Could this be an early philosophical understanding of black holes containing the seeds of new universes?

2. The Vishnu Purana's Description of Cosmic Rebirth

- The Vishnu Purana describes how at the end of one cycle, the universe collapses into an unmanifest state before being reborn.
- This echoes modern Big Bounce and black hole universe models, where the end of one universe may trigger the birth of another.

3. The Shiva Linga as a Symbol of Cosmic Cycles

- Some interpretations of the Shiva Linga suggest it represents a black hole-like entity—a singularity from which creation emerges.
- Shiva, as the cosmic destroyer and renewer, embodies the dual role of black holes—destroying matter but potentially giving rise to new realms.

These parallels suggest that black holes may not just be points of destruction but could function as Hiranyagarbha-like portals, where old universes collapse and new ones emerge.

Conclusion: Black Holes as Modern Cosmic Wombs

The idea that black holes could seed new universes has striking philosophical and scientific implications:

- In modern physics, black holes are being explored as possible gateways to new realms—mirroring the Vedic idea of cosmic rebirth through Hiranyagarbha.
- If every black hole contains the potential for a new universe, then creation is not a singular event but a continuous, self-replicating process.
- Hindu cosmology's vision of repeated cosmic cycles aligns with the idea that each black hole singularity could birth a

new reality.

If our universe was born from a black hole in another universe, and our own black holes are seeding new cosmoses, then existence itself is infinite, without a true beginning or end.

In the next section, we will explore the deeper implications of this cycle—does the universe undergo endless cycles of birth, destruction, and rebirth? If so, what does this mean for our understanding of time, existence, and our place in the cosmos?

6.4 The Cyclical Universe: From Expansion to Collapse and Rebirth

One of the most profound ideas in both Hindu cosmology and modern astrophysics is that the universe is not a one-time event but follows a continuous cycle of birth, expansion, collapse, and rebirth.

Hindu philosophy has long described the cosmos as an eternal, self-renewing entity, governed by vast cycles known as Kalpas and Yugas. Similarly, modern physics is now exploring models such as the Big Crunch, the Big Bounce, and the oscillating universe hypothesis, all of which suggest that the universe could collapse and be reborn infinitely.

Could these ancient and modern visions of cyclic time be pointing to the same fundamental truth? If the universe is truly cyclical, does this mean existence itself is endless, with no ultimate beginning or final end?

The Hindu Concept of Kalpas and Yugas—Cyclic Time and Repeated Creation

Hindu cosmology is unique in that it does not view time as linear but as cyclic, governed by vast cosmic ages that repeat endlessly. This concept is embodied in Kalpas and Yugas, which define the rise, decline, and renewal of the universe over incomprehensibly vast timescales.

1. The Cycle of Kalpas (Cosmic Days and Nights of Brahma)

- A Kalpa (day of Brahma) lasts 4.32 billion years, during which the universe is created, sustained, and eventually destroyed.
- This is followed by a cosmic night of equal duration, where the universe remains in an unmanifest state before rebirth.
- After 100 Brahma years, even Brahma dissolves into Maha Pralaya, a final cosmic dissolution before the next cosmic cycle begins.

2. The Four Yugas (Ages of the Universe)

- Each Kalpa is divided into four Yugas—Satya, Treta, Dwapara, and Kali—representing a gradual decline from perfection to chaos.
- After Kali Yuga (the current age), the cycle resets, bringing forth a new golden age (Satya Yuga).
- This idea mirrors the scientific model of an expanding universe followed by a contraction and a subsequent rebirth.

Thus, Hindu cosmology envisions the universe as undergoing endless cycles of creation and dissolution, rather than a single creation event.

Scientific Models of the Big Crunch, Big Bounce, and the Oscillating Universe

Modern physics is now exploring whether the universe might also follow a cyclical pattern, with its fate linked to the fundamental forces of gravity, dark energy, and entropy.

1. The Big Crunch Hypothesis: A Scientific Pralaya?

- If the expansion of the universe slows down and reverses due to gravitational pull, it would result in the Big Crunch—a scenario where the universe collapses back into a singularity.
- This would be remarkably similar to Hindu descriptions of Pralaya, where the universe is absorbed into Brahman (the

unmanifest cosmic reality) before the next cycle begins.

2. The Big Bounce: A Scientific Hiranyagarbha?

- Instead of ending in total collapse, some models suggest that a new universe could emerge from the remnants of the previous one.
- This is called the Big Bounce, where a collapsing universe "rebounds" into a new expansion—just as Hiranyagarbha (the Golden Womb) gives birth to the cosmos after each dissolution.

3. The Oscillating Universe Model: An Infinite Cosmic Cycle

- Some physicists propose that the universe may follow an endless cycle of expansion and contraction, much like Hindu cosmology's repeating Kalpas.
- If true, then the universe has no ultimate beginning or final end—it simply evolves through infinite cycles.

These scientific models parallel the Hindu idea that the universe is destroyed and reborn endlessly, rather than having a singular moment of creation.

Does the Universe Undergo Endless Cycles of Birth, Destruction, and Rebirth?

If both Hindu cosmology and modern physics point toward a cyclical universe, it raises profound questions:

1. Is the end of the universe simply a transition to its next form?

- In Hindu thought, Pralaya is not the end but a phase of cosmic transformation, preparing for the next cycle.
- If the Big Crunch or Big Bounce is correct, then the collapse of one universe might be the seed for the next.

2. Is time itself cyclic?

- Hinduism suggests that time is not linear but moves in vast loops, just as cosmic cycles repeat.
- Could time, like space, be subject to expansion, contraction, and rebirth?

3. Does consciousness persist through cosmic cycles?

- Hindu philosophy posits that Brahman—the eternal consciousness—remains beyond these cycles.
- If universes are reborn, could consciousness also be eternal, persisting beyond time and space?

Conclusion: The Universe as an Endless Cycle of Creation and Renewal

The convergence of ancient wisdom and modern science suggests that:

- The universe may be cyclical, undergoing infinite expansions and contractions, just as Hindu cosmology describes Kalpas and Yugas.
- Scientific models such as the Big Crunch and Big Bounce resemble Hindu descriptions of Pralaya and cosmic rebirth.
- If time and space can reset, then existence itself is not bound by a single beginning or end—it is part of an eternal, evolving reality.

If these insights hold true, then our universe is not a one-time creation but a single iteration in an infinite chain of cosmic rebirths.

In the next section, we will explore a deeper question: Does consciousness play a role in the universe's cycles? If the cosmos is reborn endlessly, does awareness itself persist across these cycles, guiding the rhythm of creation? ◈

6.5 Does Consciousness Play a Role in the Universe's Rebirth?

As we explore the cyclic nature of the universe, one fundamental question emerges: Does consciousness persist through these cycles, shaping and influencing cosmic rebirth?

Hindu philosophy describes Brahman—the ultimate, eternal reality—as an observer that transcends creation and dissolution. While the universe undergoes infinite cycles of birth, expansion, collapse, and rebirth, Brahman remains the unchanging, infinite consciousness behind all existence.

Modern physics, particularly in the realm of quantum mechanics, is beginning to entertain a similar idea: that consciousness may be fundamental to reality, shaping the very structure of the cosmos. Could it be that the act of observation, awareness, or intelligence plays a role in the unfolding and rebirth of the universe itself?

The Ancient Perspective: Brahman as the Eternal Observer of Cosmic Cycles

In Hindu cosmology, Brahman is not a deity but the ultimate reality—a boundless, timeless consciousness that exists beyond space and time. Unlike the physical universe, which undergoes creation and destruction, Brahman remains unchanged, witnessing these cosmic cycles without being affected by them.

1. Brahman and the Illusion of Reality

- The Upanishads describe the material universe as Maya, an illusion projected by Brahman's awareness.
- This suggests that the universe is not entirely independent but emerges from and dissolves back into a greater consciousness.

2. The Observer Beyond Time

- Hindu texts state that even after Pralaya (cosmic dissolution), Brahman remains as pure awareness.
- This aligns with the idea that consciousness is not bound by physical time—it exists beyond the rise and fall of universes.

3. Consciousness as the Seed of Creation

- Just as Hiranyagarbha (the Golden Womb) is the cosmic seed of creation, could it be that consciousness itself is the "first cause" of each reborn universe?
- If Brahman is the ultimate witness, then does it also guide the transition from cosmic dissolution to rebirth?

This ancient perspective suggests that consciousness is not merely a product of the universe—it may be its foundation, influencing how cosmic cycles unfold.

Could Consciousness Be a Fundamental Force, Influencing Cosmic Evolution?

In modern science, consciousness is largely considered a product of brain activity, emerging from complex neural interactions. However, some leading-edge theories in physics and philosophy propose that consciousness may be more fundamental than matter itself.

1. The Conscious Universe Hypothesis

- Some scientists argue that consciousness may not be a byproduct of physical processes but a fundamental aspect of reality.
- This aligns with the Hindu idea that consciousness (Chaitanya) is the root of all creation.

2. The Anthropic Principle: Does the Universe Need Conscious Observers?

- The Anthropic Principle suggests that the universe's physical laws seem fine-tuned to allow for conscious life.
- If the universe is reborn infinitely, could consciousness play a role in shaping which physical laws emerge in each new cycle?

3. Does Consciousness Guide Cosmic Evolution?

- If awareness persists beyond death, could it also persist beyond cosmic dissolution?
- Hindu texts imply that cosmic intelligence—Brahman—guides creation, much like information and structure seem to persist even after black hole singularities.

This raises a profound question: Does consciousness exist independently of the material universe, influencing its cycles of birth and rebirth?

Does the Act of Observation Play a Role in the Birth of New Realities, as Quantum Physics Suggests?

Quantum physics has revealed strange and counterintuitive truths about reality, particularly regarding the role of observation in shaping existence.

1. The Observer Effect: Consciousness Shapes Reality

- In quantum experiments, particles exist in multiple possible states until observed.
- This suggests that the act of observation "collapses" possibilities into a definite reality.
- Could this principle apply on a cosmic scale, influencing the way universes emerge and evolve?

2. The Wheeler's Participatory Universe Hypothesis

- Physicist John Wheeler proposed that the universe does not exist independently—it requires observers to bring it into reality.
- If true, then consciousness may be essential to the unfolding of cosmic cycles.
- This aligns with Hindu thought, where Brahman's awareness gives rise to material existence.

3. Are Universes Born Through Conscious Awareness?

- If observation shapes physical reality, could the awareness of a higher intelligence play a role in birthing universes?
- Some speculative theories suggest that black holes, quantum fluctuations, or vacuum states might require some form of information or awareness to initiate cosmic creation.

This brings us to a powerful realization: If consciousness plays a role in shaping physical reality at the quantum level, could it also influence the rebirth of entire universes?

Conclusion: Consciousness as the Cosmic Witness and Architect

Both ancient Hindu philosophy and modern physics are converging toward a revolutionary idea—that consciousness is not an incidental product of the universe but may be deeply connected to its structure and evolution.

- Hindu cosmology describes Brahman as the eternal consciousness that witnesses and transcends cosmic cycles.
- Modern physics suggests that observation plays a key role in defining reality, raising the question of whether awareness is fundamental to the universe's rebirth.
- If consciousness is eternal, could it persist beyond cosmic

dissolution, guiding the emergence of new universes in an infinite cycle?

If these insights hold true, then the universe is not just a mechanical system but a dynamic, evolving structure, deeply connected to the presence of intelligence and awareness.

7: The Chavan's Limits – Why Black Holes Are Not Singularities

In the previous chapter, we explored the possibility that consciousness plays a role in shaping the cycles of the universe, influencing cosmic rebirth rather than simply being a passive observer. But if the universe is an evolving, dynamic system, can black holes truly be absolute endpoints where time and space collapse into nothingness?

For years, the idea of singularities—points of infinite density where physical laws break down—has been a cornerstone of black hole physics. However, both quantum mechanics and modern astrophysics suggest that singularities may not actually exist in the way we once thought. Instead of being cosmic dead ends, black holes might be highly structured energy states that regulate the flow of mass, energy, and information.

This perspective aligns remarkably well with ancient Hindu cosmology, which views the universe as an ever-transforming system rather than a series of absolute beginnings and endings. Instead of collapse into an infinite point, energy undergoes constant transformation, balance, and redistribution.

In this chapter, we introduce **Universal Energy Dynamics (UED)**, a conceptual framework that redefines black holes as structured equilibrium states rather than singularities. We will explore how this model challenges conventional physics while finding deep resonance with ancient cosmological views of infinite transformation. Could it be that black holes are not paradoxes, but instead function as energy regulators maintaining universal balance—just as the cosmos has always been described in Vedic traditions?

7.1 The Flawed Concept of Singularities in Modern Physics

For decades, the idea of singularities—points of infinite density and zero volume—has been an accepted consequence of Einstein's General Theory of Relativity. According to traditional models, when a massive star collapses under its own gravity, it forms a black hole, where all matter is crushed into a single point of infinite density. This is the so-called gravitational singularity, a region where space-time curvature becomes infinite, and known laws of physics cease to function.

However, the very idea of infinity in a physical system presents severe paradoxes. No physical theory can handle true infinities—whether in energy, density, or temperature—without breaking down. This suggests that singularities, as currently understood, may not actually exist in nature. Instead, black holes may represent extreme but structured states of energy equilibrium, requiring an alternative explanation beyond classical physics.

Why the Idea of Infinite Density at a Singularity Creates Mathematical Paradoxes

The concept of a point of infinite density is not just physically problematic—it is mathematically inconsistent. Here's why:

1. Breakdown of Physical Laws

- The singularity equation suggests that density approaches infinity as volume shrinks to zero.
- This leads to undefined results in physics, meaning the singularity is not a real physical entity but a mathematical artifact.

2. The Information Paradox

- According to classical physics, any information (mass, charge,

angular momentum) that falls into a black hole is lost forever.
- However, quantum mechanics insists that information cannot be destroyed.
- If singularities were real, they would violate this fundamental principle, creating an unresolved paradox.

3. The Cosmic Censorship Hypothesis

- Physicist Roger Penrose proposed that "naked singularities" (ones exposed without an event horizon) should not exist in nature.
- If singularities were true physical points, they would cause catastrophic breakdowns in predictability, making reality fundamentally unstable.

These inconsistencies strongly suggest that singularities are not physical objects but indicators that our current understanding of gravity is incomplete.

The Limitations of Einstein's General Relativity in Extreme Gravitational Fields

While Einstein's General Theory of Relativity remains one of the most successful scientific theories of all time, it does not provide a complete picture of black hole interiors.

1. Relativity Predicts Singularities, But Cannot Handle Them

- Einstein's equations predict the existence of singularities but offer no explanation for what happens inside them.
- This means that General Relativity is incomplete in describing extreme gravitational conditions.

2. Failure at the Quantum Scale

- General Relativity breaks down at very small distances, where

quantum effects become dominant.
- To fully understand black holes, quantum gravity—a theory that merges General Relativity with quantum mechanics—is needed.

3. Event Horizons vs. Quantum Interactions

- Relativity assumes that black hole event horizons are absolute one-way barriers—nothing escapes.
- However, quantum mechanics suggests that particles can tunnel through barriers, meaning event horizons may not be as "absolute" as previously thought.

Thus, Einstein's equations are powerful for understanding large-scale cosmic structures but fail at explaining what happens inside black holes.

How Quantum Physics Contradicts the Notion of Singularities, Suggesting an Alternative Structure

Modern physics, particularly quantum mechanics and quantum field theory, provides compelling reasons why black holes cannot collapse into true singularities. Instead, they may stabilize into highly dense but structured energy states.

1. Hawking Radiation and Black Hole Evaporation

- Stephen Hawking's discovery of Hawking radiation suggests that black holes slowly lose energy over time.
- If singularities truly had infinite density, they would not allow quantum fluctuations and radiation to escape.
- This means that black holes must have an internal structure that prevents infinite collapse.

2. Quantum Gravity Theories: Singularities Do Not Exist

- Several leading-edge models suggest that black holes never actually reach a singularity but instead form ultra-dense, exotic states.
- Loop Quantum Gravity (LQG) proposes that spacetime is discrete at very small scales, preventing infinite collapse.
- String Theory suggests that singularities are replaced by "fuzzballs" of vibrating strings, preserving information.

3. Black Holes as "Planck Stars" Instead of Singularities

- Some physicists, such as Carlo Rovelli, propose that inside a black hole, collapsing matter reaches a critical density and "bounces" back outward.
- This could mean that black holes are not dead zones but energy regulators, maintaining equilibrium rather than reaching infinity.

These ideas challenge the traditional singularity concept, instead painting a picture of black holes as complex, structured entities that obey new physical laws.

Conclusion: The Need for a New Model Beyond Singularities

The evidence from mathematics, relativity, and quantum physics strongly suggests that black holes are not singularities but highly organized energy systems.

- Singularities lead to mathematical paradoxes and breakdowns in physics.
- Einstein's General Relativity predicts singularities but cannot explain them.
- Quantum mechanics indicates that singularities do not truly exist and must be replaced by a structured alternative.

This leads us to **Universal Energy Dynamics (UED)**, a framework that redefines black holes as structured equilibrium states, balancing mass, energy, and information instead of collapsing into infinite points. In the next section, we introduce UED and explore how black holes function as structured energy hubs rather than destructive singularities.

7.2 Universal Energy Dynamics (UED): A New Paradigm for Black Holes

With the breakdown of the singularity concept in modern physics, we must look for a new framework that explains black holes not as paradoxical infinities but as structured energy states. This is where Universal Energy Dynamics (UED) comes into play—a model that redefines black holes as energy regulators, balancing cosmic forces rather than serving as ultimate points of collapse.

UED suggests that black holes are not singularities but structured states of equilibrium, where energy and mass are neither infinitely compressed nor permanently lost. Instead, energy flows, transforms, and redistributes, ensuring the continued balance of cosmic structures.

The Principles of UED—Why Black Holes Are Structured States of Energy, Not Singularities

Traditional physics treats black holes as regions where all matter collapses to a singularity of infinite density. However, UED proposes a different perspective:

1. Black Holes Are Structured Energy States, Not Infinite Points

- Instead of singularities, black holes stabilize into a state of maximum energy compression without reaching infinity.
- Just as a neutron star stabilizes at an extreme density instead of collapsing into nothingness, black holes may reach a similar structured state.

2. Gravity and Energy Maintain an Internal Equilibrium

- Gravity pulls matter inward, but quantum effects, space-time geometry, and energy redistribution prevent infinite collapse.
- This means black holes are dynamic entities where forces interact in a way that maintains a structured balance.

3. The Interior of a Black Hole May Contain a New State of Matter or Energy

- If UED is correct, then inside black holes, energy is compressed into a state that we do not yet fully understand—but it is not an infinite singularity.
- This could be a new form of exotic energy, quantum vacuum structure, or an ultra-dense phase of space-time.

In short, UED suggests that black holes are not breakdowns of physics but highly sophisticated cosmic mechanisms that regulate energy, much like stars regulate fusion.

How Energy Flows and Transforms Rather Than Collapsing Into an Absolute Point

In a singularity model, all mass and energy collapse into a zero-volume point, violating fundamental conservation laws. However, UED proposes that black holes instead act as cosmic transformers, where energy follows a structured pathway of redistribution.

1. Energy Compression Instead of Infinite Collapse

- UED suggests that black holes reach a threshold where further compression is resisted by quantum effects.
- Similar to how electron degeneracy prevents white dwarfs from collapsing further, or neutron degeneracy stops neutron stars from infinite collapse, UED proposes that another mechanism prevents singularities.

2. Energy May Flow In and Out of Black Holes in Ways We Do Not Yet Fully Understand

- Hawking radiation already suggests that energy escapes black holes over time.
- Some models propose that black holes may not be one-way traps but instead interact with space-time in a way that redistributes energy to the universe.

3. Does Energy Transform Inside Black Holes?

- If energy does not infinitely collapse, it must transform into another state.
- This could mean that inside black holes, energy is reorganized into a new phase—possibly linked to the quantum vacuum, holographic principles, or alternate dimensions.

In this view, black holes are not objects of destruction but part of a grand cosmic cycle where energy is conserved, transformed, and redistributed.

Why Black Holes May Act as Hubs of Energy Balance Rather Than Regions of Destruction

Under the traditional model, black holes are seen as regions where matter and energy disappear forever. But under UED, black holes function more like hubs of energy balance, regulating cosmic structures rather than simply erasing information.

1. Cosmic Stability and the Role of Black Holes

- Galaxies revolve around supermassive black holes at their centers.
- If black holes were purely destructive, they would destabilize cosmic structures rather than support them.
- Instead, they appear to play a key role in maintaining galaxy

formation and evolution.

2. Black Holes as Energy Regulators

- Rather than being bottomless pits of gravity, black holes may act as self-contained energy structures that interact dynamically with space-time.
- They might store, transform, and eventually release energy in ways we do not yet fully understand.

3. The Cosmic Recycling of Energy

- If black holes are not singularities, then the energy that enters them is not lost but instead converted into a different form.
- This could mean that black holes participate in cosmic recycling, ensuring that energy remains part of the universe's evolution.

These ideas align strikingly well with ancient cosmological perspectives, which emphasize transformation rather than absolute endings.

Conclusion: Black Holes as Dynamic Energy Structures, Not Singularities

The Universal Energy Dynamics (UED) model provides an alternative to the flawed singularity concept, presenting black holes as structured hubs of energy regulation rather than infinite collapse points.

- Singularities violate known physical principles, but UED suggests that black holes stabilize into structured equilibrium states.
- Energy does not vanish inside black holes; instead, it transforms and redistributes in unknown but structured ways.
- Black holes may function as energy regulators within the

cosmic ecosystem, much like ancient cosmology describes transformation instead of destruction.

If UED is correct, then black holes are not the end of physics but rather gateways to a deeper understanding of energy and space-time.

In the next section, we will explore how black holes maintain energy equilibrium and what this means for the structure of space-time itself. ◈

7.3 The Structure of Energy Equilibrium in Black Holes

If black holes are not singularities but structured states of energy equilibrium, as Universal Energy Dynamics (UED) suggests, then their role in the cosmos must be reconsidered. Rather than acting as one-way gravitational traps where matter and energy disappear forever, black holes may function as cosmic energy regulators—absorbing, transforming, and even redistributing energy back into the universe.

This perspective aligns with quantum physics and modern black hole research, which suggests that energy is not truly lost but undergoes transformation within black holes. Could black holes be self-contained energy systems that maintain a form of internal balance rather than collapsing into nothingness? In this section, we explore how black holes may redistribute energy, maintain stable energy cores, and function as part of a larger cosmic equilibrium system.

How Black Holes May Redistribute Energy Rather Than "Trap" It Forever

The traditional view of black holes as regions where nothing escapes has been increasingly challenged by modern physics. The discovery of Hawking radiation, information conservation theories, and quantum gravity models suggests that black holes do not simply "trap" energy but actively interact with the universe.

1. Hawking Radiation and Energy Release Over Time

- According to Stephen Hawking, black holes emit radiation due to quantum effects near the event horizon.
- This means that black holes are not completely isolated—they lose mass and energy over time.
- If this process continues long enough, a black hole could eventually evaporate completely, returning all of its energy to the universe.

2. The Role of Black Holes in Galactic Formation

- Supermassive black holes sit at the center of most galaxies, playing a role in their stability and evolution.
- Rather than acting as purely destructive entities, black holes seem to interact with surrounding matter, influencing star formation, gas distribution, and galactic structure.
- This suggests that black holes are active participants in cosmic energy redistribution.

3. Theoretical Models of Energy Escape Mechanisms

- Some theories propose that energy absorbed by black holes is later released in unexpected ways—such as through white holes, Hawking radiation, or quantum gravitational effects.
- This means that black holes may function more like energy transformers rather than final endpoints.

In this view, black holes do not "trap" energy permanently but act as regulatory systems that process and eventually return energy to the universe.

The Possible Existence of Stable Energy Cores Instead of Singularities

If black holes do not collapse into singularities, then what exists at their core? UED proposes that instead of an infinite-density point, black holes may contain a stable energy core—a region where gravitational and quantum forces reach a balance, preventing further collapse.

1. Quantum Effects Prevent Total Collapse

- Quantum gravity suggests that spacetime may have a fundamental limit to how much it can be compressed.
- If this is true, then inside black holes, matter does not collapse into a singularity but instead reaches a maximum compression point.
- This could result in a stable energy core where gravity, quantum effects, and space-time structure interact dynamically.

2. The "Planck Star" Hypothesis

- Some physicists propose that instead of a singularity, black holes stabilize into "Planck stars"—ultra-dense energy states that maintain a structured equilibrium.
- If true, this would mean that black holes do not lead to absolute destruction but instead exist as stable, energy-dense objects that retain their structure over time.

3. The Role of Vacuum Energy and Space-Time Foam

- In quantum physics, empty space is not truly empty—it is filled with vacuum fluctuations and quantum energy.
- Some theories propose that at the heart of a black hole, vacuum energy interactions prevent infinite collapse, creating a self-sustaining energy core.

If these ideas are correct, then black holes are not singularities but cosmic regulators that reach a structured equilibrium rather than collapsing into infinite density.

Could Hawking Radiation and Black Hole Evaporation Be Part of a Larger Energy Balance System?

Hawking radiation suggests that black holes slowly release energy, meaning they are not entirely isolated systems. But could this process be part of a much larger cosmic energy balance mechanism?

1. Black Holes as Temporary Energy Reservoirs

- If energy is absorbed into black holes and then gradually released through Hawking radiation, this means black holes act as temporary storage systems rather than permanent traps.
- This aligns with the idea that energy cycles through different forms rather than disappearing.

2. The Connection Between Black Hole Evaporation and Cosmic Expansion

- Some models suggest that energy released from black holes over billions of years may contribute to cosmic expansion.
- This means that rather than being separate entities, black holes may play an integral role in the evolution of the universe itself.

3. Black Holes as Energy Equilibrium Systems Rather Than Destructive Singularities

- If black holes absorb, store, and eventually return energy, they may act as part of a larger energy equilibrium network across the universe.
- This idea aligns with ancient cosmological concepts of transformation and balance, rather than absolute destruction.

Conclusion: Black Holes as Energy Regulators, Not Absolute Endpoints

The Universal Energy Dynamics (UED) model challenges the traditional notion of singularities and instead proposes that black holes function as structured energy systems that regulate, transform, and redistribute energy rather than destroying it.

- Hawking radiation suggests that energy slowly escapes black holes, meaning they are not permanent energy traps.
- The possible existence of stable energy cores instead of singularities suggests that black holes may reach equilibrium rather than infinite collapse.
- Black holes may function as cosmic energy regulators, ensuring the transformation and balance of energy rather than leading to total destruction.

If UED is correct, then black holes are not paradoxes but key components of a cosmic energy balance system that aligns with both modern physics and ancient cosmological perspectives.
In the next section, we will explore how ancient cosmology's vision of infinite transformation aligns with modern views of black holes as structured, evolving entities. ◈

7.4 Ancient Cosmology's View on Infinite Transformation

As modern physics moves away from the idea of black hole singularities as absolute endpoints, it begins to align with ancient cosmological views that describe the universe as an ever-evolving system of transformation, rather than a series of ultimate collapses.
Hindu and Vedic cosmology do not envision the cosmos as a one-time event leading to a final collapse. Instead, they describe reality as a continuous process of formation, dissolution, and renewal. This idea resonates with Universal Energy Dynamics (UED), which suggests that black holes, rather than being one-way traps, act as energy regulators, balancing and redistributing cosmic forces.

Could it be that ancient wisdom already understood the principle of infinite transformation, which modern physics is only now beginning to recognize? If so, Shiva's Tandava, the balance of Brahman, and the Chavan's Limits may provide a deeper insight into how black holes function—not as singularities but as structured systems of cosmic equilibrium.

Hindu and Vedic Cosmology's Vision of Constant Transformation Rather Than Absolute Collapse

In Hindu cosmology, the universe is not seen as a static or linear system with a beginning and an end. Instead, it follows an eternal cycle of creation, preservation, and transformation.

1. The Concept of Cyclical Universes (Kalpas and Yugas)

- The universe undergoes periodic dissolution (Pralaya) but is always reborn, continuing an infinite cycle.
- Each cycle is governed by energy balance rather than absolute destruction.

2. Pralaya as Transformation, Not Annihilation

- The universe dissolves into a dormant but structured state before re-expanding.
- This is similar to how black holes absorb energy but later release it, maintaining cosmic equilibrium.

3. Energy Flow Instead of Finality

- Ancient cosmology suggests that nothing truly disappears; energy and matter only change forms.
- This aligns with UED's idea that black holes regulate energy rather than serving as dead ends.

Thus, ancient cosmology rejects the idea of infinite collapse, instead describing the universe as an interconnected, evolving system of transformation—an idea now supported by modern physics.

The Concept of Shiva's Tandava (Cosmic Dance) as an Energy Equilibrium, Not Destruction

One of the most profound representations of universal transformation in Hinduism is Shiva's Tandava—the cosmic dance of creation and dissolution. This dance symbolizes the balance between expansion and contraction, the continuous movement of energy, and the interplay of forces that sustain the universe.

1. Shiva as Nataraja: The Lord of Cosmic Balance

- Shiva's dance is not about destruction alone—it represents rhythmic transformation, ensuring continuity rather than total annihilation.
- This mirrors the function of black holes as cosmic balancers, rather than objects of infinite collapse.

2. The Tandava's Two Aspects: Creation and Dissolution

- Ananda Tandava (Dance of Bliss) represents expansion and creative forces.
- Rudra Tandava (Fierce Dance) represents dissolution—but not absolute destruction.
- Both exist in balance, just as energy in black holes follows structured transformation rather than collapse.

3. Shiva's Dance and Universal Energy Dynamics (UED)

- The Tandava mirrors the idea that energy is never lost but only moves through different states.
- This is consistent with modern views that black holes do not trap energy forever but participate in energy redistribution.

Thus, Shiva's cosmic dance serves as an ancient metaphor for what modern physics now suggests: energy is never destroyed but continually transformed in a structured process.

How the Chavan's Limits Align with the Principle of Brahman—The Eternal Transformation of Reality

The Chavan's Limits challenge the classical notion of black hole singularities by proposing that black holes reach an equilibrium state where energy transformation continues, rather than collapsing into nothingness. This concept finds a striking parallel in the Hindu philosophical concept of Brahman, the eternal, unchanging essence that underlies all reality.

1. Brahman as the Foundation of Transformation

- According to Vedanta philosophy, Brahman is neither created nor destroyed—it simply exists as the underlying energy of the cosmos.
- If black holes act as cosmic balancers rather than singularities, they too function as part of this eternal transformation process.

2. The Chavan's Limits and the Dynamic Universe

- If energy is never truly lost, then black holes must operate under an equilibrium principle, where mass-energy is cycled rather than infinitely compressed.
- This aligns with the idea of Brahman as a constant source that allows transformation without absolute loss.

3. The Role of Perceived Endings in Cosmic Evolution

- Just as black holes appear to be endpoints but actually regulate cosmic energy, ancient texts describe Pralaya (dissolution) as a transition rather than finality.

- This suggests that the perceived "end" of a black hole could actually be the beginning of a new state of existence—aligning with the Hindu vision of infinite cosmic renewal.

Thus, both UED and ancient cosmology reject the notion of absolute singularities, instead viewing transformation as an eternal, structured process.

Conclusion: Ancient Wisdom and Modern Physics Converge

The idea of infinite collapse is not supported by either ancient cosmology or modern physics. Instead, black holes—and the universe itself—operate on principles of energy equilibrium, transformation, and cyclical balance.

- Hindu cosmology describes the universe as an ever-evolving, balanced system, not an absolute collapse.
- Shiva's Tandava represents structured transformation, mirroring the role of black holes in maintaining cosmic energy balance.
- The Chavan's Limits align with the principle of Brahman, reinforcing the idea that energy is never lost, only transformed.

If black holes are not singularities but structured systems of energy redistribution, then we must rethink how they fit into the grand design of the universe. Could black holes be functioning as cosmic processors, regulating the very fabric of space-time rather than tearing it apart?

In the next section, we will explore how this understanding of black holes as structured energy systems may reshape the future of physics and our understanding of reality itself. ◈

7.5 Beyond Singularities: The Future of Black Hole Physics

As our understanding of black holes evolves, the traditional singularity model is being questioned by both theoretical physics and emerging scientific discoveries. The idea that black holes are merely gravitational traps leading to infinite collapse is proving to be incomplete, if not entirely flawed. Instead, new research suggests that black holes could be structured, dynamic energy systems rather than ultimate endpoints.

Universal Energy Dynamics (UED) challenges the notion that black holes are cosmic dead zones. Instead, it proposes that black holes function as energy regulators, redistributing mass and energy rather than collapsing into singularities. This shift in perspective could revolutionize our understanding of space-time, energy flow, and the fundamental nature of the universe itself.

In this section, we explore why black hole research is moving beyond singularities, how UED could reshape space-time physics, and whether black holes serve as structured energy gateways rather than cosmic endpoints.

Why Black Hole Research Is Shifting Away from the Singularity Model

For decades, black holes were thought to contain singularities—regions of infinite density where space-time collapses entirely. However, several major developments have cast doubt on this assumption.

1. Singularities Are Mathematically Problematic

- Einstein's General Relativity predicts singularities, but they create paradoxes where known physics breaks down.
- Infinite density and zero volume are not physically meaningful concepts, suggesting that the singularity model is incomplete.

2. Quantum Mechanics and Gravity Do Not Support Singularities

- Quantum physics suggests that energy and information cannot be destroyed, contradicting the idea of a singularity absorbing everything without releasing it.
- The Hawking Radiation paradox hints that black holes must have an internal structure capable of processing and eventually releasing energy.

3. New Theoretical Models Provide Alternative Explanations

- Loop Quantum Gravity (LQG) proposes that extreme gravitational collapse stops at a Planck-scale energy state, preventing singularities.
- The Firewall Hypothesis suggests that instead of singularities, black holes may have energetic barriers or transitional layers where quantum effects dominate.
- Fuzzball Theory (from String Theory) suggests that black holes are not points of infinite density but highly structured "fuzzy" energy formations.

These developments indicate that black holes may not be singularities but instead complex energy structures that interact with space-time in previously unknown ways.

How UED Could Redefine Space-Time Physics and the Role of Black Holes

Universal Energy Dynamics (UED) provides a new framework for understanding black holes as structured, self-regulating energy states rather than paradoxical singularities.

1. Black Holes as Structured Energy Nodes

- Instead of collapsing into infinite density, UED suggests that black holes reach a stable equilibrium, where energy transformation takes place.

- These structures may resemble ultra-dense cores of energy rather than singularities.

2. Energy Redistribution and Space-Time Interaction

- UED posits that black holes do not simply absorb energy but also redistribute it through space-time.
- This would explain how black holes influence their surrounding galaxies, regulate cosmic structures, and even contribute to universal expansion.

3. The Role of Black Holes in Maintaining Space-Time Integrity

- If black holes are energy processors rather than dead ends, they could play a key role in shaping and stabilizing space-time itself.
- This aligns with ideas in quantum gravity, which propose that black holes are woven into the fundamental structure of the universe, rather than breaking it apart.

By shifting away from the singularity model, UED suggests that black holes are not "bottomless pits" but rather structured energy hubs that operate as part of a larger cosmic system.

Could Black Holes Be Gateways to Structured Energy Flows Rather Than Cosmic Endpoints?

One of the most exciting possibilities emerging from the rejection of the singularity model is the idea that black holes could be transition points between different states of energy and space-time.

1. Black Holes as Cosmic Transformers

- If black holes absorb and redistribute energy, they might serve as cosmic transformers, regulating mass-energy balance across

the universe.
- This could explain why galaxies form around black holes and why they appear to play a stabilizing role in the cosmos.

2. The White Hole Hypothesis and Energy Conservation

- Some models suggest that black holes might be connected to white holes—regions where absorbed energy is released into new space-time realms.
- If true, this means that black holes are not true endpoints but energy pathways.

3. Black Holes and the Concept of Wormholes

- Some researchers propose that black holes could act as entry points into higher-dimensional space or parallel universes.
- This aligns with theories in String Theory and M-Theory, which suggest that black holes could be bridges between different layers of reality.

If black holes are not final states of collapse but structured energy flows, they could be fundamental to the evolution of the universe, acting as cosmic engines rather than mere gravitational traps.

Conclusion: The Future of Black Hole Physics and the End of the Singularity Myth

The future of black hole physics no longer revolves around the outdated singularity model. Instead, new theories—ranging from quantum gravity to UED—suggest that black holes are dynamic, structured entities that regulate energy and space-time.

- Singularities are being replaced by models of structured energy states, where gravitational collapse is balanced by quantum effects.

- UED proposes that black holes function as cosmic energy regulators, shaping galaxies and contributing to space-time evolution.
- Instead of being gravitational endpoints, black holes may serve as structured energy flows, possibly acting as gateways between cosmic regions or dimensions.

If these ideas hold true, then black holes are not paradoxes but essential components of the universe's balance, transforming and redistributing energy in ways we are only beginning to understand.

8: The Hidden Order – Dark Energy and the Unseen Forces Governing the Universe

In the previous chapter, we explored how black holes may not be singularities but structured energy regulators, maintaining cosmic equilibrium rather than serving as endpoints of gravitational collapse. However, black holes are not the only mystery governing the universe's fate—another enigmatic force dominates the cosmos: dark energy.

Modern physics tells us that dark energy accounts for nearly 70% of the total energy in the universe, yet we know almost nothing about its nature. It is believed to be responsible for the accelerating expansion of space-time, pushing galaxies apart and counteracting the pull of gravity. But instead of being an unknown, repulsive force, what if dark energy is part of a structured energy redistribution system—an unseen yet fundamental mechanism that sustains the universe?

Surprisingly, ancient cosmology describes a similar concept of an all-pervading cosmic energy known as Prana, the force that drives motion, sustains balance, and permeates all of existence. Could Prana be an early philosophical insight into the very force modern science is now trying to define? If so, does this mean that dark energy is not an inexplicable anomaly but an essential component of the universe's hidden order?

In this chapter, we explore how dark energy might not be a force of chaos but rather a mechanism of structured energy flow, how ancient descriptions of Prana align with modern theories of vacuum energy, and whether these ideas can be integrated to redefine our understanding of space-time itself. Could dark energy be the key to understanding the true nature of cosmic expansion, the role of black holes, and the fundamental equilibrium governing the universe?

8.1 The Mystery of Dark Energy: A Force That Defies Understanding

Dark energy is one of the greatest cosmic enigmas of modern physics. First detected through its unexpected influence on the expansion of the universe, it accounts for nearly 70% of all the energy present in the cosmos, yet we have no clear understanding of what it actually is. Unlike dark matter, which exerts a gravitational pull, dark energy appears to act in the opposite way—pushing space apart, accelerating the universe's expansion.

When scientists first discovered this phenomenon, it challenged our fundamental understanding of gravity and the dynamics of space-time. Instead of slowing down due to gravitational attraction, the universe was found to be expanding at an increasing rate, defying Newtonian and relativistic expectations. This discovery raised profound questions:

- Is dark energy a new, undiscovered fundamental force?
- Does it represent an intrinsic property of space-time itself?
- Or could it be a manifestation of a deeper, structured energy redistribution system that governs cosmic equilibrium?

These questions have yet to be fully answered, but emerging theories suggest that dark energy might not be a new force at all—but rather an effect of deeper energy interactions at the fabric of reality.

How Dark Energy Was Discovered and Why It Contradicts Existing Models of Gravity

For most of human history, scientists assumed that the expansion of the universe would eventually slow down due to the gravitational pull of all the matter within it. However, in the late 1990s, astronomers analyzing distant supernovae found that the expansion of the universe is not decelerating—it is accelerating.

1. The Supernova Surveys

- By measuring the light from Type Ia supernovae, which serve as standard candles for distance measurements, astronomers expected to confirm a gradual slowing down of cosmic expansion.
- Instead, they found that these supernovae were farther away than expected, meaning that the universe's expansion was speeding up rather than slowing down.

2. The Challenge to General Relativity

- According to Einstein's General Theory of Relativity, gravity should cause large-scale structures to pull toward each other, slowing cosmic expansion over time.
- The discovery of accelerated expansion meant that something was counteracting gravity—an unseen force driving space apart.

3. Einstein's "Biggest Blunder" May Have Been Correct

- Einstein had originally introduced a cosmological constant (Λ) to balance gravity, ensuring a static universe.
- After Hubble's discovery of cosmic expansion, Einstein removed this term, calling it his biggest blunder.
- However, the discovery of dark energy suggests that some form of Λ-like energy is indeed responsible for cosmic acceleration.

This acceleration defies the conventional understanding of how gravity works on a cosmic scale, suggesting that either: A new force is at play, or Energy redistribution is occurring in ways we do not yet understand.

The Problem with Treating Dark Energy as an "Unknown Repulsive Force"

The standard explanation for dark energy is that it behaves like a repulsive force, counteracting gravity on the largest scales. However, this idea presents several challenges:

1. No Clear Mechanism for Repulsion

- Unlike other fundamental forces (gravity, electromagnetism, strong and weak nuclear forces), dark energy has no clear mechanism or force carrier.
- If dark energy is truly a force, why does it not interact with matter in any direct, observable way?

2. Why Does Dark Energy Remain Constant?

- The energy density of dark energy appears to be constant per unit volume, unlike other known forms of energy.
- This contradicts many existing models, making dark energy anomalous compared to known physical laws.

3. Is It Just an Effect of Space Itself?

- Some physicists propose that dark energy is not a force at all but an intrinsic property of space-time.
- This means that rather than being a "new" force, dark energy could be a result of the natural expansion and redistribution of energy across the universe.

If dark energy is not an independent force, then what is it? Could it be an emergent phenomenon resulting from deeper space-time interactions, similar to how Prana in Vedic cosmology is described as an all-pervading energy that sustains balance?

Why Dark Energy Might Be an Effect of Structured Energy Redistribution Rather Than a New Fundamental Force

Emerging theories suggest that dark energy may not be a separate force but rather a form of structured energy redistribution within the universe. This aligns with the idea that energy in the universe is not randomly scattered but follows certain patterns and flows that maintain cosmic equilibrium.

1. Vacuum Energy and Quantum Fluctuations

- Quantum field theory suggests that even empty space contains energy in the form of vacuum fluctuations.
- Some physicists propose that dark energy arises from the quantum vacuum—meaning it is not a new force but an effect of the fundamental nature of space-time itself.

2. A Self-Regulating Cosmic Energy System

- If energy cannot be created or destroyed, then dark energy may be part of a self-regulating system that maintains large-scale balance.
- Instead of thinking of dark energy as an active "push" force, it could be a natural result of the universe maintaining equilibrium at the largest scales.

3. The Connection to Black Holes and Cosmic Voids

- Some theories propose that dark energy is influenced by the behavior of black holes and cosmic voids, which act as energy redistribution centers.

- Rather than being separate forces, dark energy and gravity may be two sides of the same underlying energy system—similar to how Prana is seen as a single cosmic energy that flows through all things.

This interpretation of dark energy as a structured energy flow rather than a mysterious force could help bridge the gap between modern physics and ancient cosmological insights, which describe energy as an interconnected, self-balancing phenomenon rather than an independent force.

Conclusion: Rethinking Dark Energy as a Hidden Order

The discovery of dark energy challenged the foundations of physics, forcing scientists to rethink the nature of gravity, expansion, and cosmic equilibrium. However, treating dark energy as an unknown repulsive force may not be the right approach.

- Dark energy does not behave like a traditional force—it may instead be an effect of deeper space-time properties.
- If energy follows structured redistribution patterns, dark energy could be part of a larger cosmic balance rather than an anomaly.
- This concept aligns with ancient ideas of Prana—an all-pervading energy flow that sustains and balances the cosmos.

If dark energy is indeed a structured energy flow rather than an unknown force, then understanding its nature may reveal the true structure of space-time and its relationship with gravity, black holes, and the expansion of the universe.

In the next section, we will explore how dark energy interacts with cosmic structures and whether its effects can be understood as part of a deeper universal energy system. �

8.2 The Role of Dark Energy in Cosmic Expansion

Dark energy is often described as the invisible force that accelerates the expansion of the universe, but this explanation is incomplete. Unlike gravity, which attracts and pulls matter together, dark energy appears to do the opposite—causing galaxies to move farther apart at an increasing rate. However, this raises a fundamental question:

Is dark energy truly a separate force, or is it simply an effect of a deeper, structured energy mechanism within space-time itself?

Rather than being an independent, repulsive force, emerging theories suggest that dark energy could be an intrinsic property of space-time—a manifestation of a self-regulating energy equilibrium that governs large-scale cosmic expansion. If true, this would mean that dark energy is not working against gravity but rather coexisting with it in a hidden balance.

This section explores how dark energy shapes the motion of galaxies, whether it functions as an intrinsic part of space-time, and how it might interact with black holes and other cosmic structures to maintain equilibrium.

How Dark Energy Governs the Large-Scale Motion of Galaxies and Space-Time

For most of the 20th century, scientists expected that the gravitational pull of galaxies and dark matter would eventually slow the expansion of the universe. Instead, they found that the rate of expansion is accelerating, meaning some unknown factor is overpowering gravity on cosmic scales.

1. Galactic Drift and the Expanding Cosmos

- Observations of distant galaxies show that the farther a galaxy is, the faster it moves away from us.
- This suggests that the very fabric of space-time is expanding,

not just individual galaxies moving apart.

2. The Scale at Which Dark Energy Becomes Dominant

- At small scales (within galaxies and clusters), gravity dominates, holding matter together.
- At large scales (beyond galactic clusters), dark energy takes over, pushing space apart.
- This implies that dark energy is not a uniform force but an effect that operates based on scale.

3. Does Space-Time Have an Internal Expansion Mechanism?

- If dark energy is not a new force but an inherent property of space-time, then expansion may be a natural consequence of how energy is distributed across the universe.
- This suggests that the universe expands not because of an external force but because of an internal energy dynamic.

In this view, dark energy does not push galaxies apart as an independent force but instead results from the way space-time itself maintains balance.

The Idea That Dark Energy Is Not a Separate Force but a Manifestation of Deeper Space-Time Properties

Rather than treating dark energy as a mysterious anti-gravitational force, some scientists propose that it is an emergent phenomenon—a natural feature of space-time itself.

1. The Cosmological Constant as a Property of Space

- Einstein's cosmological constant (Λ) was originally proposed as a built-in energy density of space itself.
- This idea suggests that dark energy is not a separate entity but an inherent feature of the universe's fabric.

2. The Zero-Point Energy of Quantum Space

- In quantum mechanics, even "empty" space contains energy due to constant quantum fluctuations.
- Dark energy could be an effect of this underlying quantum vacuum energy.

3. A Self-Regulating Expansion Mechanism?

- If dark energy is simply a balancing mechanism of space-time, then expansion occurs as part of a structured energy redistribution system.
- This would mean that dark energy is not competing with gravity but maintaining a universal balance.

If true, this means that dark energy is not an unknown force but a structured component of the universe, ensuring that energy remains evenly distributed across vast cosmic scales.

Could Dark Energy Interact with Black Holes and Other Cosmic Structures to Maintain Equilibrium?

If dark energy is a fundamental aspect of space-time, it should naturally interact with other cosmic structures, including black holes, galaxies, and cosmic voids.

1. Black Holes as Energy Sinks and Cosmic Stabilizers

- While dark energy pushes space apart, black holes pull matter inward, creating a balancing effect.
- If black holes regulate energy within galaxies, could they also interact with dark energy to maintain cosmic equilibrium?

2. The Expansion of Cosmic Voids

- Some research suggests that dark energy is strongest in cosmic voids—regions of near-empty space between galaxy clusters.

- If dark energy is an effect of space-time rather than a force, then its presence in voids could indicate a natural redistribution of energy in low-density regions.

3. Are Black Holes and Dark Energy Two Sides of the Same Mechanism?

- Black holes absorb and transform energy at extreme densities.
- Dark energy appears where mass is low and space-time stretches outward.
- This suggests that black holes may act as counterbalances to dark energy, keeping energy in structured equilibrium across the universe.

This interpretation aligns with the idea that energy in the universe is not chaotic but follows structured pathways, maintaining cosmic balance rather than simply expanding indefinitely.

Conclusion: Dark Energy as a Structured Feature of Space-Time, Not an Anomaly

Rather than being a mysterious force that contradicts gravity, dark energy may be a manifestation of deeper space-time properties—an effect of structured energy redistribution rather than an independent force.

- Dark energy governs large-scale expansion, but it does not act uniformly—it appears strongest in cosmic voids and weakest in dense regions.
- Rather than opposing gravity, dark energy may be part of a larger energy equilibrium, maintaining balance across cosmic scales.
- Its interaction with black holes and cosmic voids suggests that

dark energy is not a chaotic force but part of a structured energy regulation mechanism.

If this is true, then understanding dark energy requires rethinking the universe as a system of dynamic, structured energy flows rather than isolated forces acting independently.

In the next section, we will explore whether ancient cosmology's concept of Prana—the all-pervading life energy—could provide insights into the true nature of dark energy. Could the Vedic concept of Prana be an early intuition of what modern science now calls dark energy? ⟐

8.3 Ancient Cosmology and the Concept of an All-Pervading Energy (Prana)

As modern physics struggles to define dark energy, ancient cosmology may already provide an intuitive understanding of this mysterious force. In Hindu and Vedic philosophy, Prana is described as the all-pervading cosmic energy that sustains motion, maintains balance, and drives transformation across the universe. This concept, deeply embedded in ancient texts, closely mirrors modern discussions on vacuum energy, dark energy, and the structured flow of cosmic forces.

Could it be that dark energy is not a new discovery but rather a modern rediscovery of what ancient civilizations already understood—an intrinsic, dynamic energy system that governs the universe's expansion and balance? If so, does this mean that Prana and dark energy are two different ways of describing the same fundamental principle?

In this section, we explore how the Vedic concept of Prana aligns with modern discussions on vacuum energy, how ancient wisdom saw energy as a self-regulating system rather than a chaotic force, and why the idea that energy is never lost but always transformed resonates with both ancient philosophy and modern physics.

The Vedic Concept of Prana as a Universal Energy That Sustains and Drives Motion

In Vedic tradition, Prana is the fundamental life force that animates all existence, from the motion of celestial bodies to the breath within living beings. It is considered:

- Eternal and infinite, neither created nor destroyed but constantly flowing through different forms.
- Omnipresent, pervading the cosmos at every scale, from subatomic particles to galaxies.
- A dynamic force, responsible for all movement, transformation, and balance in the universe.

This idea closely resembles the role dark energy plays in modern astrophysics, as an unseen force influencing the expansion, motion, and distribution of energy across cosmic structures.
Prana's Role in the Universe According to Ancient Texts

- In the Upanishads, Prana is described as the breath of the cosmos, the sustaining force that moves all things.
- In Yoga and Ayurveda, Prana is not only associated with life but also with the movement of energy in space and time.
- The Rig Veda speaks of an omnipresent energy that connects the macrocosm (the universe) with the microcosm (individual existence).

These descriptions suggest that Prana is more than just a life force—it is an all-encompassing energy that organizes the flow of existence. This view aligns closely with modern discussions of dark energy as an unseen force shaping the expansion of the universe.

How Ancient Descriptions of an Omnipresent Energy Field Resemble Modern Discussions of Vacuum Energy and Dark Energy

Modern physics describes space not as empty, but as filled with quantum energy fluctuations, known as vacuum energy. This idea is strikingly similar to the Vedic concept of an omnipresent, invisible energy (Prana) permeating all of existence.

1. The Vacuum Energy of Modern Physics

- Quantum mechanics reveals that even in a perfect vacuum, energy fluctuations continuously occur.
- These fluctuations create a background energy field that exists everywhere, even in empty space.
- This is similar to how Prana is described as an energy field that sustains all movement and change.

2. Dark Energy as a Manifestation of Vacuum Energy

- Some physicists believe that dark energy is an effect of vacuum energy—a property of space itself rather than a separate force.
- This aligns with the Vedic view that energy is woven into the very fabric of existence, sustaining cosmic processes without needing a distinct origin.

3. The Cosmic Web and Structured Energy Flows

- The large-scale structure of the universe reveals that galaxies and matter form a vast cosmic web, connected by unseen forces.
- Ancient cosmology described the universe as a structured energy system, where Prana flows through interconnected layers, just as modern science now describes energy flowing

through cosmic structures.

These parallels suggest that ancient wisdom and modern physics may be describing the same fundamental phenomenon using different language and perspectives.

The Philosophical Idea That Energy Is Never Lost but Always Cycling Through Different States—Mirroring Modern Conservation Laws

A key principle in both ancient cosmology and modern physics is that energy is never truly lost—it only changes form.

1. The Law of Conservation of Energy

- In physics, the First Law of Thermodynamics states that energy cannot be created or destroyed, only transformed.
- This is remarkably similar to the Vedic concept that Prana is eternal, shifting between different states but never ceasing to exist.

2. Prana as a Cyclic Energy System

- Ancient texts describe Prana as flowing through cycles—creation, preservation, and transformation—much like the universe's expansion and contraction.
- This aligns with the idea that dark energy may not be an independent force but part of a larger, structured energy cycle that governs the universe.

3. The Role of Black Holes in Energy Transformation

- Just as Prana cycles through different states, modern physics suggests that black holes do not simply destroy matter and energy but redistribute them in new forms.
- If dark energy interacts with black holes and other cosmic

structures, it could be part of a grand cosmic balance rather than an isolated force.

These insights suggest that ancient cosmology's emphasis on an eternal, self-regulating energy field closely mirrors modern physics' shift toward understanding the deeper structure of space-time and energy redistribution.

Conclusion: Prana and Dark Energy as Two Perspectives of the Same Cosmic Principle

The concept of Prana in ancient Vedic philosophy and the modern mystery of dark energy share remarkable similarities:

- Both describe an invisible, omnipresent energy that sustains movement and cosmic balance.
- Both suggest that energy is never lost but cycles through different states.
- Both align with the idea that space is not empty but filled with an unseen force that governs large-scale motion.

If dark energy is truly a self-regulating energy field rather than a chaotic force, then ancient cosmology may have already provided a conceptual framework for understanding its nature. This raises profound questions:

- Is dark energy an integral feature of a cosmic intelligence that self-regulates the universe?
- Could modern physics learn from ancient insights to develop a more complete theory of energy and space-time?
- If energy is structured rather than random, could dark energy be an active participant in cosmic equilibrium rather than a passive anomaly?

These questions lead us to the next section, where we explore whether modern physics can integrate these ideas to reshape our understanding of space-time, and what this means for the future of cosmology. ◈

8.4 Can Modern Physics Integrate These Ideas to Reshape Our Understanding of Space-Time?

As we have seen, both ancient cosmology and modern physics recognize the existence of an all-pervading energy—whether it is called Prana in Vedic philosophy or dark energy in astrophysics. The challenge now is to integrate these insights into a unified framework that explains space-time, cosmic expansion, and energy equilibrium.

Emerging theories such as Quantum Field Theory (QFT) and Zero-Point Energy suggest that the universe is not an empty void but filled with unseen energy fluctuations. This aligns with the idea that dark energy may not be a mysterious repulsive force but a dynamic aspect of space-time itself. If true, then dark energy is not an anomaly—it is part of the very structure of reality, influencing the balance of the cosmos, interacting with gravitational fields, and shaping cosmic evolution.

In this section, we explore how modern physics is beginning to embrace these ideas, how dark energy might be better understood as a property of space-time rather than an unknown force, and how it could be linked to cosmic structures like black holes and cosmic voids.

How Emerging Theories Like Quantum Field Theory and Zero-Point Energy Suggest That Space Is Filled with Unseen Energy

For centuries, scientists believed that space was an empty void, a mere backdrop for cosmic events. However, modern physics has shattered this idea, revealing that space itself is a dynamic entity filled with fluctuating energy fields.

1. Quantum Field Theory (QFT): Space Is Never Truly Empty

- In QFT, all of reality is described as a set of interacting quantum fields that extend through space-time.
- Even in a vacuum, these fields are constantly fluctuating, meaning that space is never truly empty—it is filled with virtual particles and energy.
- This aligns with the ancient idea that energy is omnipresent, flowing through all things, even where we perceive emptiness.

2. Zero-Point Energy: The Energy of the Vacuum

- Quantum physics suggests that even in a perfect vacuum, there is still residual energy known as zero-point energy.
- This energy is not directly observable, but its effects can be measured (such as in the Casimir effect, where vacuum fluctuations exert a small force on objects).
- If dark energy is linked to zero-point energy, it could mean that the accelerating expansion of the universe is not caused by a separate force but by the underlying fluctuations of space itself.

3. The Energy Structure of Space-Time

- In General Relativity, space-time is not static—it bends, stretches, and interacts with mass and energy.
- If space-time itself carries an intrinsic energy field, dark energy could be a manifestation of how this field evolves on cosmic scales.
- This would mean that dark energy is not "acting" on the universe—it is part of the universe's natural dynamics.

This idea directly aligns with the ancient Vedic concept that energy is everywhere, flowing in structured ways to maintain balance, much like Prana does within the cosmos.

Does Dark Energy Represent the Dynamic Energy Fluctuations of Space-Time Rather Than a Separate Force?

Rather than being a new, unexplained force, dark energy may simply be an effect of how energy interacts with space-time.

1. A Self-Regulating Energy Field

- If space-time has an intrinsic energy component, then its expansion is not driven by an external force but by a self-regulating mechanism that distributes energy over time.
- This would mean that dark energy is not an independent force but a natural consequence of space-time's properties.

2. Fluctuations at Cosmic Scales

- At the quantum level, space-time exhibits small-scale fluctuations due to vacuum energy.
- On a cosmic scale, these fluctuations could manifest as large-scale expansion, appearing as dark energy.
- This would explain why dark energy is strongest in cosmic voids, where gravitational effects are weakest.

3. From Expansion to Contraction?

- Some theories suggest that if dark energy is a space-time fluctuation effect, it may not lead to eternal expansion but instead could reverse over time, leading to a new cosmic cycle.
- This aligns with Vedic cosmology, where Prana is cyclical—expanding and contracting through different

cosmic phases.

If dark energy is a manifestation of space-time itself, rather than a separate force, it fundamentally changes how we view the universe—not as a place where forces act independently but as a self-organizing system where energy moves through structured cycles.

Could Dark Energy Be Linked to Gravitational Balance and Cosmic Structures Like Black Holes and Cosmic Voids?

If dark energy is not an external force but a structured energy system within space-time, then it should interact with cosmic structures such as black holes, galaxies, and voids.

1. Black Holes as Energy Regulators

- Black holes absorb matter and energy, while dark energy spreads energy across space-time.
- This suggests that black holes and dark energy could be working in balance, regulating the flow of energy in the universe.
- If energy is never lost but always transformed, then black holes may serve as counterbalancing structures to dark energy's expansive force.

2. Cosmic Voids and Dark Energy Concentration

- Observations show that dark energy effects are strongest in cosmic voids—regions with little to no matter.
- This suggests that dark energy is not uniform but concentrated in areas where gravitational forces are weak.
- If true, then dark energy may be part of a larger mechanism that redistributes energy across the universe to maintain equilibrium.

3. Dark Energy as a Gravity-Compensating Mechanism

- Gravity pulls objects together, but if space-time has an intrinsic energy component, it may compensate for gravity's effects by expanding.
- This suggests that dark energy and gravity are not opposing forces but two aspects of a deeper cosmic balance.

If dark energy is part of a structured energy regulation system rather than a chaotic expansion force, it could mean that the universe is self-organizing, with energy cycling through different phases rather than expanding infinitely.

Conclusion: Toward a Unified Understanding of Space-Time and Cosmic Energy

Modern physics is beginning to uncover what ancient cosmology has long suggested—that space is not empty, but filled with a dynamic, structured energy system that governs the universe's evolution.

- Quantum Field Theory and Zero-Point Energy suggest that space is filled with energy fluctuations, aligning with the Vedic idea of Prana as an all-pervading force.
- Dark energy may not be a separate force but an effect of space-time's intrinsic energy dynamics.
- Black holes and cosmic voids may interact with dark energy as part of a larger gravitational and energy balance system.

If these ideas hold true, then dark energy is not a random expansion mechanism—it is part of a grand cosmic equilibrium that ensures the universe's stability.

This raises profound questions:

- Is the universe fine-tuned through structured energy cycles rather than random expansion?

- Do black holes and dark energy function as regulators in an interconnected cosmic balance?
- Can we develop a new space-time model that integrates both ancient insights and modern physics to fully explain cosmic energy dynamics?

In the next section, we explore the future of space-time physics and how these ideas may lead to a new understanding of the universe—not as a chaotic expanding void, but as an evolving, self-regulating system of structured energy flow. ◈

8.5 The Future of Space-Time Physics: A Unified View of Energy and Expansion

For centuries, humanity's understanding of the universe has evolved from a static, unchanging cosmos to a dynamic and expanding space-time. However, the discovery of dark energy and the accelerating expansion of the universe has challenged our fundamental understanding of gravity, energy, and the structure of reality itself.

The search for dark energy is not just about identifying a missing force—it is forcing a complete rethinking of space-time physics. If dark energy is not a separate force but a structured equilibrium process within space-time, it could lead to a paradigm shift in how we view cosmic energy, black holes, and universal evolution.

In this section, we explore why the search for dark energy could redefine space-time, how a structured energy balance might replace the chaotic expansion model, and whether bridging ancient and modern insights could lead to a more complete theory of cosmic energy.

Why the Search for Dark Energy Might Lead to a Paradigm Shift in Our Understanding of Space-Time

Dark energy was originally introduced as a placeholder—an unknown force added to equations to explain an unexpected phenomenon. However, emerging theories suggest that dark energy is not an anomaly, but a clue pointing toward a deeper reality that modern physics has yet to fully comprehend.

1. Beyond the Expanding Universe Model

- For decades, we have assumed that space-time simply expands due to an unknown force.
- But if dark energy is a structured process rather than a random force, then space-time may behave in ways we do not yet understand.
- This could mean that expansion is not just a one-way trajectory but a cyclic or self-regulating mechanism.

2. Space-Time as an Active System

- Instead of being a passive backdrop, space-time may be an active participant in energy balance.
- If dark energy is an inherent property of space, then it is not working against gravity, but coexisting with it in a deeper, hidden order.
- This challenges the traditional model of physics where forces act separately, and instead suggests an interconnected energy system.

3. A New Theory of Space-Time?

- If dark energy is part of a self-regulating energy balance, we may need a new model of space-time that integrates expansion, gravity, and quantum fluctuations into a single framework.
- This could lead to an alternative to both the singularity model and the infinite expansion model, offering a more structured, evolving view of the universe.

These possibilities suggest that dark energy is not just an unexplained force—it is a sign that our entire understanding of cosmic energy must evolve.

The Idea That Energy in the Universe Follows a Structured Equilibrium Rather Than Acting as a Chaotic Force

Modern cosmology often describes the universe as expanding due to a repulsive force (dark energy), with no clear underlying order. However, if dark energy is part of a structured equilibrium, then energy distribution across the cosmos is not random—it follows a deeper pattern.

1. The Balance Between Expansion and Contraction

- If black holes and dark energy are two sides of a cosmic energy balance, then the universe is not expanding uncontrollably.
- Instead, expansion and contraction work together, ensuring a structured equilibrium that prevents chaos.

2. The Role of Cosmic Webs and Voids

- Observations show that galaxies are arranged in vast cosmic webs, with dark energy appearing stronger in cosmic voids.
- This suggests that energy is not distributed randomly, but follows structured pathways that guide the evolution of space-time.

3. From Infinite Expansion to Energy Regulation

- If the universe follows a structured energy cycle, then expansion may not continue indefinitely.
- Instead, energy may be redistributed through cosmic mechanisms, maintaining balance rather than leading to

infinite dispersion.

This structured energy model closely resembles ancient cosmological views, which describe cosmic energy as flowing in an ordered system rather than acting as a chaotic force.

Can We Bridge Ancient and Modern Insights to Develop a More Complete Theory of Cosmic Energy?

While modern physics is now exploring ideas that ancient cosmology proposed thousands of years ago, there is still a divide between scientific models and philosophical insights. However, emerging evidence suggests that combining these perspectives could lead to a more holistic understanding of cosmic energy.

1. The Vedic Concept of Prana as a Cosmic Energy Flow

- In Vedic tradition, Prana is not just life energy—it is the fundamental force that sustains movement and transformation.
- This idea resembles modern views of space-time energy fluctuations, suggesting a deeper unity between ancient philosophy and modern science.

2. Dark Energy as the Evolution of Cosmic Intelligence?

- If dark energy is a structured mechanism of space-time balance, then the universe is not a chaotic system—it is an evolving, self-regulating entity.
- Could this mean that energy itself follows a form of intelligence, adapting to cosmic structures in a way that ensures long-term equilibrium?
- This resonates with ancient descriptions of the universe as a conscious, evolving system.

3. A New Space-Time Model: Structured Energy Cycles Instead of Random Expansion

- Modern physics may eventually replace the idea of a purely expanding universe with a model that incorporates cycles, balance, and energy flow.
- This would align with both quantum physics and Vedic cosmology, leading to a unified understanding of how cosmic energy evolves over time.

By bridging the ancient and modern perspectives on energy, we may be able to construct a more complete theory of space-time—one that not only explains dark energy but redefines the way we understand the universe itself.

Conclusion: The Future of Space-Time Physics and the Next Cosmic Revolution

The discovery of dark energy has forced a fundamental shift in cosmology, requiring us to rethink our models of space, energy, and time.

- Rather than being a chaotic force, dark energy may be part of a structured energy system that balances cosmic evolution.
- The idea that space-time itself has intrinsic energy aligns with ancient cosmological insights, suggesting a hidden order in the universe.
- A new framework of cosmic energy may emerge—one that unifies black holes, gravitational equilibrium, and dark energy into a self-regulating system.

If these ideas hold true, then the universe is not an aimless expansion, but an evolving, structured entity governed by intelligent energy flows. This raises profound questions for the future of physics:

- Can we build a unified theory that integrates dark energy, gravity, and quantum fields into a single energy framework?
- Does energy in the universe follow cycles of expansion, redistribution, and balance rather than infinite expansion?
- Is space-time itself a form of intelligent energy, guiding cosmic evolution rather than acting as a passive medium?

As we move forward, these ideas may lead to a new scientific revolution—one where physics and cosmology merge into a holistic understanding of the universe as a structured, self-sustaining energy system.

9: The Universe's Final Breath – The Cycle of Expansion and Contraction

In the previous chapter, we explored how dark energy might be an intrinsic property of space-time, ensuring a structured energy balance rather than a chaotic force driving the universe apart. But if the universe expands in a controlled manner, does this expansion continue forever, or does it eventually reverse into a contraction, leading to a new cycle of cosmic rebirth?

Ancient Indian cosmology describes a cyclical universe, where cosmic creation and dissolution occur repeatedly in vast timescales, resembling the breathing of the cosmos—expanding during Brahma's day and contracting during Brahma's night. This idea closely parallels emerging scientific theories that propose the Big Crunch or Big Bounce—a possible end to cosmic expansion, followed by a contraction that seeds the birth of a new universe.

But what mechanism could drive this transition? Could black holes, the mysterious gravitational wells at the heart of galaxies, hold the key to a self-renewing cosmic process? Rather than being endpoints of destruction, could they be gateways that recycle energy, ensuring the continual rebirth of the cosmos?

This chapter explores the possibility that the universe follows a grand cycle of expansion and contraction, the role of black holes in this transformation, and how these ideas could redefine the future of astrophysics and humanity's role in understanding cosmic evolution.

9.1 Brahma's Day and Night: The Ancient Concept of Cosmic Cycles

Across ancient civilizations, the nature of time and the cosmos has been one of the greatest mysteries. While modern science largely follows a linear view of time, ancient Hindu cosmology presents a radically different perspective—a universe that breathes, expanding and contracting in grand cycles.

In Hindu philosophy, the universe is not a single event with a definitive beginning and end; rather, it follows a rhythmic cycle of creation, preservation, and dissolution, known as Kalpas. These cycles, often described as Brahma's day and night, offer a vision of the cosmos that closely resembles modern theories of a cyclic universe, where expansion is eventually reversed, leading to a cosmic rebirth.

Could this ancient framework align with scientific models that propose the universe undergoes continuous phases of expansion and contraction, like the breathing of a cosmic entity? If so, does this mean that the universe is not moving toward eternal heat death but is part of an infinite loop of renewal?

The Hindu Concept of Kalpas – Cosmic Time Cycles of Creation and Dissolution

Hindu cosmology describes time on an inconceivably vast scale, measured in Kalpas, which are long cosmic cycles consisting of repeated phases of creation and dissolution. According to the Puranas and Vedic texts, these cycles follow a structured process:

1. Brahma's Day (Creation and Expansion)

- A Kalpa begins with the birth of the universe, known as Srishti (creation).
- During this time, matter, space, and energy expand, forming galaxies, stars, and planetary systems.
- This phase mirrors modern cosmology's description of the Big Bang and cosmic expansion.

2. Brahma's Night (Contraction and Dissolution)

- As Brahma's day ends, the universe gradually contracts and dissolves.
- All matter collapses into a dormant state, and the universe enters Pralaya (cosmic dissolution).
- This phase bears a striking resemblance to the Big Crunch or Big Bounce hypothesis.

3. The Cycle Resets

- After Pralaya, a new Kalpa begins, marking the rebirth of the universe.
- This endless cycle continues, ensuring that creation and destruction are two sides of the same cosmic reality.

This cyclical vision of the cosmos directly challenges the Western idea of a universe with a single beginning and an eventual end. Instead, it suggests that existence is eternal, moving through continuous phases rather than ceasing to exist.

How Brahma's Day (Expansion) and Night (Contraction) Resemble Modern Theories of Cyclic Universes

Modern cosmology has long debated the fate of the universe. While early theories suggested that the universe would expand indefinitely until it became cold and lifeless (heat death), newer models propose that the universe may follow cycles of expansion and contraction, resembling the Kalpa cycles described in ancient texts.

1. The Big Bang and Cosmic Expansion (Brahma's Day)

- The Big Bang theory describes how the universe emerged from an extremely hot and dense state, expanding outward to form galaxies, stars, and planetary systems.
- This process mirrors Brahma's day, where creation unfolds, and the universe expands into existence.

2. The Big Crunch and Cosmic Contraction (Brahma's Night)

- Some theories suggest that gravity could eventually overcome expansion, causing the universe to contract.
- Matter and energy would collapse back into a singular state, much like Pralaya—the cosmic dissolution described in Hindu texts.

3. The Big Bounce – A Cycle of Infinite Renewal

- If contraction reaches a critical density, it may lead to a new cosmic explosion, birthing another universe.
- This model, called the Big Bounce, suggests that the universe expands, contracts, and then expands again, forming an eternal cycle.
- Such a process would mean that our universe is not the first, nor the last, but part of an infinite chain of cosmic rebirths.

These parallels suggest that the ancient concept of Kalpas and Brahma's day and night may offer an early intuitive understanding of the cyclic nature of the cosmos.

Could the Universe Follow a Grand Cosmic Breath—Expanding and Contracting in an Eternal Cycle?

One of the most poetic descriptions of the universe in Hindu cosmology is that it breathes—expanding and contracting like a living entity. This aligns with modern cyclic universe models that propose expansion and contraction occur as natural processes, much like the inhalation and exhalation of a breath.

1. Space-Time as a Dynamic, Living System

- If space-time is not static but behaves like a living entity, then expansion and contraction could be natural aspects of its existence.

- Instead of an infinite, cold void, the universe would be a self-renewing structure that continuously transforms itself.

2. Energy Conservation in a Cyclic Universe

- According to the laws of thermodynamics, energy cannot be created or destroyed—only transformed.
- This suggests that even in cosmic dissolution, energy is not lost but reorganized, ready to fuel the next cycle of creation.
- Could black holes serve as the mechanism through which energy is recycled between cosmic cycles?

3. The Illusion of an Ending

- If cosmic time follows cycles, then what we perceive as an end (the universe's heat death or collapse) may simply be a transition into the next phase.
- This perspective challenges the Western notion of a final cosmic fate and instead embraces an eternal, evolving cosmos.

Conclusion: The Universe as an Eternal, Self-Renewing System

The Hindu concept of Kalpas and Brahma's day and night offers a vision of the universe as a continuously evolving cycle rather than a linear trajectory toward an ultimate end.

- The expansion of the universe (Brahma's Day) mirrors modern cosmology's understanding of cosmic inflation and the Big Bang.
- The eventual contraction of the universe (Brahma's Night) resembles the Big Crunch and cyclic universe models.
- The idea of the universe as a grand breath—a continuous cycle of creation and dissolution—aligns with emerging theories in

physics that suggest the universe is self-renewing rather than finite.

If the universe does breathe through cycles of expansion and contraction, then black holes, gravitational forces, and space-time itself may be deeply involved in regulating this cosmic balance.
This leads us to our next section:

- What role do black holes play in this cosmic renewal?
- Could they serve as bridges between cycles, preserving the energy and information needed for the next universe?
- Do they hold the key to understanding how the universe resets itself?

In the next section, we will explore how black holes may act as the ultimate cosmic regulators, ensuring that energy and matter are never lost but simply transformed, sustaining the universe's eternal cycle.

9.2 The Fate of the Universe: Infinite Expansion or Eventual Contraction?

The ultimate fate of the universe remains one of the most profound unanswered questions in cosmology. While ancient Hindu texts describe cosmic cycles of expansion and dissolution, modern physics is still debating whether the universe will expand forever, tear itself apart, or eventually collapse back into a singular state.
Current astrophysical models propose three possible endings:

- Heat Death (The Universe Freezes to Stillness)
- The Big Rip (The Universe Tears Itself Apart)
- The Big Crunch (The Universe Collapses and Rebirths)

Each of these scenarios depends on the role of dark energy, gravity, and black holes in shaping cosmic evolution. But which of these outcomes

is most likely? Could black holes and gravitational forces eventually reverse expansion, leading to a rebirth?
This section explores the three leading theories of cosmic fate, why an infinitely expanding universe creates paradoxes, and whether gravity and black holes hold the key to a cyclic cosmic renewal.

The Three Competing Theories of the Universe's Fate

Scientists currently propose three major possibilities for the future of our universe. Each outcome depends on how dark energy behaves and whether gravity can eventually slow or reverse cosmic expansion.

1. Heat Death – A Universe That Freezes to Stillness

This theory, also called the Big Freeze, is based on the assumption that dark energy will continue to accelerate expansion forever.

- As galaxies drift farther apart, stars will burn out, black holes will slowly evaporate (via Hawking radiation), and all energy will be evenly spread out.
- Without new sources of heat, the universe will become a cold, dark, lifeless void—a cosmic wasteland.
- This would mark the end of all structure, energy flow, and motion in the cosmos.

Why Heat Death Presents a Paradox:

- What happens to the fundamental energy of the universe? The laws of physics state that energy cannot be created or destroyed—so where does it go?
- Could the "dead" universe still contain potential for renewal? Some physicists propose that even in the coldest, most desolate universe, quantum fluctuations could eventually

trigger a new Big Bang.

2. The Big Rip – When Expansion Becomes Too Violent

In this scenario, dark energy does not just accelerate expansion—it becomes stronger over time.

- The fabric of space-time stretches at an increasing rate, pulling galaxies, stars, and even atoms apart.
- Eventually, the expansion force becomes so extreme that it rips apart everything—even fundamental particles.
- The universe would disintegrate into nothingness, eliminating any chance of future structure or cycles.

Why the Big Rip Presents a Problem:

- Why would dark energy increase over time? There is currently no direct evidence that dark energy strengthens as the universe expands.
- What remains after the Rip? If the universe completely dissolves, what (if anything) could emerge next?

3. The Big Crunch – The Universe Reverses Expansion

The third possibility, and one that aligns closely with ancient cosmology, is that gravity eventually halts expansion and pulls everything back inward.

- Over billions of years, the pull of gravity may overpower dark energy, causing galaxies to slow down and reverse their outward motion.
- The universe would begin collapsing in on itself, compressing

into a singular state.
- This could lead to another Big Bang, marking the start of a new cycle—a concept known as the Big Bounce.

Why the Big Crunch Supports a Cyclic Universe:

- It aligns with fundamental energy conservation laws—matter and energy are never lost, only transformed.
- It mirrors ancient ideas of cosmic cycles—expansion and contraction as part of an eternal rhythm.

Could the Big Crunch be the modern scientific equivalent of Brahma's Night, where the universe dissolves into a singularity before its next rebirth?

Why an Infinitely Expanding Universe Creates Unsolved Paradoxes

The idea of an eternally expanding universe may seem logical, but it introduces paradoxes that physics struggles to resolve.

1. The Problem of Entropy and Energy Loss

- If energy becomes evenly distributed, how can new cosmic structures form?
- The laws of thermodynamics suggest that entropy (disorder) always increases, meaning the universe should eventually reach a state of complete inactivity.

2. The Information Paradox

- If black holes evaporate, what happens to the information they contained?
- Some theories suggest that information is never truly lost but

stored within space-time, possibly leading to future cosmic cycles.

3. Why Would the Universe Have a "One-Time" Expansion?

- If the universe was born from a singularity in the Big Bang, why should expansion be a one-time event?
- It is more natural to assume that the universe follows cycles of creation and destruction rather than expanding indefinitely.

These challenges suggest that there must be mechanisms—perhaps involving black holes or gravitational forces—that regulate expansion and contraction.

Is There Evidence That Gravity and Black Holes Could Reverse Expansion Over Time?

If the universe is to contract, something must counteract dark energy and pull matter back inward. Black holes and gravity may play a crucial role in this process.

1. Black Holes as Energy Regulators

- Black holes do not just absorb matter—they reshape space-time and influence cosmic expansion.
- Some theories suggest that black holes could serve as stabilizers, slowing down expansion over time.

2. Supermassive Black Holes at the Heart of Galaxies

- Every galaxy contains a supermassive black hole at its center.
- If black holes grow large enough, their collective gravitational pull could eventually outweigh the effects of dark energy.

3. Could Black Holes Trigger the Next Big Bang?

- Some physicists propose that black holes contain "baby universes" within them, acting as portals to new cosmic cycles.
- If true, then rather than being endpoints, black holes could be gateways to future universes.

If gravity and black holes eventually overpower dark energy, the universe could stop expanding and collapse inward, setting the stage for the next cosmic cycle. This would provide a scientific foundation for the concept of Brahma's Night—where the universe dissolves before a new creation begins.

Conclusion: Is the Universe Destined for Rebirth?

While modern physics is still uncertain about the universe's fate, there are strong indications that it may not expand forever.

- Heat Death (Big Freeze) presents paradoxes regarding energy conservation and renewal.
- The Big Rip suggests a destructive end, but it lacks strong observational support.
- The Big Crunch (leading to a Big Bounce) aligns closely with ancient cosmology and modern gravitational theories.

If the universe follows a cycle of expansion and contraction, much like the breath of a cosmic entity, then the role of black holes, gravity, and space-time energy dynamics may hold the key to understanding this cycle.

In the next section, we explore the mysterious role of black holes in cosmic renewal—are they simply the remains of dying stars, or do they hold the key to preserving information and energy for the next cosmic era? ◈

9.3 Black Holes and the Rebirth of the Cosmos

Black holes are often viewed as the ultimate endpoints of matter and energy, where everything that crosses the event horizon is lost forever. However, emerging theories suggest that black holes may not be dead ends, but rather cosmic recyclers—playing a crucial role in shaping the fate of the universe.

If the universe follows cycles of expansion and contraction, could black holes serve as transition points between these cycles? Instead of being singularities of infinite destruction, could they act as portals or cosmic seeds, preserving information and energy for the birth of a new universe?

In this section, we explore how black holes may be the missing link between cosmic death and rebirth, their role in preserving fundamental information, and whether they contain the blueprint for the next cycle of existence.

How Black Holes Could Act as Cosmic Recyclers, Preserving Energy and Information

According to classical physics, when matter and energy fall into a black hole, they are lost forever. However, modern theories challenge this idea, suggesting that black holes may not destroy information but instead store or transform it.

1. Hawking Radiation and Energy Redistribution

- Physicist Stephen Hawking proposed that black holes slowly evaporate by emitting radiation (now called Hawking radiation).
- If black holes gradually release energy back into the universe, they may serve as energy redistribution hubs rather than absolute endpoints.
- This challenges the idea that black holes permanently erase

energy and suggests they play a role in maintaining cosmic balance.

2. The Black Hole Information Paradox

- One of the biggest mysteries in physics is what happens to information that falls into a black hole.
- The laws of quantum mechanics state that information cannot be destroyed, yet black holes appear to consume everything they absorb.
- Some theories suggest that black holes encode information on their event horizon, meaning that nothing is truly lost—only transformed.
- If black holes preserve information, could they also serve as the archives of past universes, storing the fundamental data needed for the next cosmic cycle?

3. Black Holes as Cosmic Energy Converters

- Some scientists propose that black holes do not trap energy forever but instead recycle it in different forms.
- If matter and energy are compressed inside a black hole, could they be released in another form—possibly in another region of space or another universe?
- This opens the possibility that black holes are not just destructive forces but mechanisms for cosmic renewal.

The Possibility That Black Holes Are Not Endpoints, but Gateways to the Next Universe

If black holes do not simply erase matter and energy, but instead transform them into a new state, then they could act as gateways between cosmic cycles.

1. White Holes – The Other Side of a Black Hole?

- Some theories propose that black holes have a mirror counterpart known as a white hole.
- Unlike black holes, which absorb everything, white holes expel matter and energy outward, potentially seeding a new region of space.
- Could white holes be the birthplaces of new cosmic structures, acting as the other end of a black hole's gravitational tunnel?

2. The Einstein-Rosen Bridge (Wormholes and Multiverse Theories)

- Some interpretations of General Relativity suggest that black holes could be wormholes—bridges connecting different parts of the universe or even different universes.
- If true, then black holes may transfer matter and energy into entirely new realms of existence, potentially triggering new cosmic cycles.

3. Black Holes as Star Seeds for the Next Universe

- If a universe collapses into a singularity (Big Crunch), the final compressed state may form a supermassive black hole containing all the information from the previous cosmos.
- This ultra-dense singularity could explode outward in a new Big Bang, forming the next cycle of existence.
- In this scenario, our universe may have been born from the remains of a previous one, just as future universes could emerge from black holes in our cosmos.

If black holes serve as cosmic passageways rather than dead ends, then rather than being the universe's graveyards, they could be the cosmic wombs where the next reality is born.

Do Black Hole Singularities Contain the Seeds of the Next Cosmic Cycle?

The concept of a singularity—a point of infinite density at the heart of a black hole—has long troubled physicists. However, recent research suggests that singularities may not be infinite points, but rather ultra-compressed states of structured energy.

1. The Problem with Infinite Density

- The idea of a true singularity (where physics breaks down) is likely an incomplete description.
- Instead, the singularity may be a transition state where energy transforms rather than disappearing.

2. The Loop Quantum Gravity Perspective

- Some physicists propose that quantum effects prevent singularities from becoming infinite.
- Instead of collapsing into nothingness, a black hole's core may reach a compressed but finite state, from which expansion can occur again.

3. The Big Bounce and the Cosmic Recycling Hypothesis

- If singularities are not true endpoints, they may serve as seeds for the next Big Bang.
- This aligns with the Big Bounce theory, where the universe undergoes continuous cycles of collapse and rebirth.
- Could black holes contain the preserved information from past universes, guiding the formation of the next cosmic era?

If true, then every black hole could be a potential cosmic embryo, holding the blueprint for the next iteration of reality.

Conclusion: Are Black Holes the Cosmic Bridges Between Life, Death, and Rebirth?

While black holes were once considered one-way tickets to oblivion, modern research suggests that they may be the key to understanding the cyclic nature of the cosmos.

- Rather than being endpoints, black holes may be cosmic recyclers, preserving energy and information.
- They could serve as gateways, transporting energy into new regions of space or triggering future universes.
- Singularities may not be infinite dead zones but highly structured energy states, acting as the seeds for the next cosmic cycle.

If black holes do play a role in cosmic rebirth, then they may be the missing link between the dissolution of one universe and the creation of another—just as Brahma's Night gives way to a new dawn of cosmic existence.

This raises profound questions for the future of physics and philosophy:

- Do black holes store the fundamental patterns of existence, allowing cosmic cycles to continue?
- Is the next universe already forming inside the black holes of our current cosmos?
- Could black holes be harnessed in the distant future to unlock deeper secrets about reality?

In the next section, we will explore the Big Bounce hypothesis—could the universe itself be part of an eternal cycle of contraction and expansion, much like the ancient concept of Kalpas and cosmic breaths?

9.4 The Big Bounce: A Universe That Breathes Instead of Dies

For decades, scientists have debated whether the universe will expand forever, collapse into a singularity, or cycle through endless phases of destruction and rebirth. While the Big Bang theory explains how the universe began, it does not address what happens at its supposed endpoint. Will it simply fade into cold emptiness, or does it have the potential to collapse and start again?

A growing number of physicists believe in the Big Bounce hypothesis—a model in which the universe does not end but instead undergoes infinite cycles of expansion and contraction, much like a cosmic breath.

If true, this theory aligns with ancient Hindu cosmology, which describes the universe as going through vast cycles of creation and dissolution, mirroring Brahma's day and night. Instead of a single beginning and end, the universe might be eternal, continuously regenerating itself through gravitational collapse and rebirth.

In this section, we explore the Big Bounce hypothesis, the elasticity of space-time, and how quantum gravity may prevent total collapse, ensuring that the universe never truly ends.

The Emerging Big Bounce Hypothesis – Does the Universe Collapse and Re-Expand?

The Big Bounce is an alternative to both the Big Bang and the Big Crunch. Instead of a universe that is born once and eventually fades into nothingness, this model proposes that the universe contracts after reaching its maximum expansion, collapses into a high-energy state, and then re-expands into a new cycle of cosmic evolution.

1. A Universe That Never Truly Ends

- Instead of a one-time explosion, the universe moves through cycles, expanding and contracting like a living system.

- Each collapse leads to a high-energy, dense state that triggers a new expansion, much like the birth of a new universe.
- This process could repeat infinitely, ensuring that the cosmos is eternal and ever-renewing.

2. The Role of Gravity in the Bounce Mechanism

- If gravity eventually overcomes expansion, the universe may begin contracting.
- As the universe shrinks, matter and energy become compressed into an ultra-dense state.
- Rather than collapsing into a singularity, some unknown mechanism—possibly quantum effects—triggers a rebound, leading to a new cycle of expansion.

3. The Evidence for Cyclic Universes

- Some cosmologists believe that traces of past universes may exist in the cosmic microwave background radiation (CMB).
- If the universe has gone through previous cycles, there could be hidden imprints of earlier cosmic epochs embedded in the structure of space-time.
- This would suggest that our universe is just the latest in an infinite series of cosmic rebirths.

This model provides a compelling alternative to the idea that the universe will either expand forever (leading to heat death) or collapse into a singularity (leading to a final end). Instead, it proposes that space-time itself has a built-in resilience, allowing it to reset and restart in a self-regulating cosmic cycle.

Could Space-Time Be Elastic, Allowing for Infinite Cycles of Rebirth?

One of the key ideas behind the Big Bounce is that space-time behaves more like a flexible, dynamic entity rather than a rigid framework.

1. A Self-Regulating Cosmic Fabric

- If space-time is elastic, it could stretch during expansion and contract during collapse, never reaching an absolute limit.
- Instead of breaking down into a singularity, the universe could undergo a smooth transition from contraction to expansion.
- This idea mirrors the concept of cosmic breath—an endless cycle of inhalation (expansion) and exhalation (contraction).

2. Quantum Foam and the Hidden Structure of Space-Time

- Some theories propose that at the smallest scales, space-time is not a smooth continuum but a fluctuating quantum foam.
- If true, this means that the fabric of reality is inherently dynamic, capable of bending and shifting rather than collapsing.
- This could explain why the universe does not break down at the moment of maximum compression, but instead rebounds into a new expansion.

3. The Expansion-Contraction Balance

- If space-time is naturally elastic, then cosmic cycles could be built into the laws of physics themselves.
- Expansion and contraction would not be opposing forces but two complementary phases of a grand cosmic process.
- This aligns with the Hindu concept of cosmic equilibrium, where creation and destruction are not opposites but essential parts of the same reality.

If the universe functions in this way, then there is no absolute end—only a continuous cycle of renewal.

The Quantum Connection – Do Quantum Gravity Effects Prevent Total Collapse?

A major challenge in understanding cosmic rebirth is what happens at the moment of maximum collapse. If the universe contracts into a singularity, why would it bounce back instead of disappearing into nothingness?

Quantum gravity offers a possible solution.

1. Loop Quantum Gravity (LQG) and the Prevention of Singularities

- In classical physics, a singularity is a point of infinite density where all known laws break down.
- However, loop quantum gravity suggests that quantum effects prevent true singularities from forming.
- Instead of collapsing infinitely, the universe would reach a highly compressed but finite state, leading to a natural rebound.

2. Planck-Scale Physics and the Cosmic Bounce

- The Planck scale is the smallest unit of measurement in quantum physics, where space-time behaves differently than in classical physics.
- Some models propose that at the Planck scale, gravity becomes repulsive rather than attractive, triggering a cosmic bounce instead of a total collapse.
- This would mean that the universe never truly reaches an absolute end—it simply transitions from one cycle to the next.

3. Implications for the Nature of Reality

- If the universe is cyclical, then time itself may not have a beginning or an end—it may be part of an eternal process.
- Instead of a random, one-time event, cosmic evolution would be part of an organized, repeating structure.
- This supports the idea that existence is not finite but infinite, mirroring the ancient Hindu concept of the eternal cosmic breath.

If quantum gravity effects prevent singularities, then black holes, cosmic collapses, and even the entire universe may be part of a never-ending loop of renewal.

Conclusion: The Universe as a Living, Breathing System

The Big Bounce hypothesis provides a radical but elegant explanation for the fate of the universe:

- Instead of expanding forever or collapsing into a final singularity, the universe follows infinite cycles of expansion and contraction.
- Space-time may be elastic, capable of stretching and contracting without reaching a true breaking point.
- Quantum gravity effects could prevent total collapse, allowing each contraction to seed a new expansion, much like a rebirth.

If this is true, then our universe is just one chapter in an eternal story—one cycle in an infinite process of cosmic renewal.
This raises profound questions:

- Has this cycle already repeated countless times before?
- Are traces of previous universes hidden in the structure of space-time?
- Is the universe itself a living entity, following a rhythm that mirrors biological life cycles?

In the next section, we explore what this means for the future of astrophysics, human understanding, and our place in this grand cosmic cycle. ◈

9.5 What This Means for the Future of Astrophysics and Human Understanding

The idea that the universe follows a cyclic pattern of expansion and contraction has profound implications—not just for astrophysics, but for our understanding of time, space, energy, and even our place in the cosmos. If the universe does not have a true beginning or an absolute end, but instead moves through endless cycles of birth, transformation, and rebirth, then reality itself may be far more dynamic than we have ever imagined.

This raises critical questions:

- What does a cyclic universe mean for the nature of time and existence?
- How does this model challenge the traditional concept of singularities and reshape space-time physics?
- Could human understanding of cosmic cycles lead to technological breakthroughs in energy, space travel, or even ways to harness the power of black holes?

In this section, we explore how a cyclic universe transforms our understanding of physics, the potential for new discoveries, and what this might mean for humanity's future.

If the Universe Is Cyclic, What Does This Mean for the Nature of Time and Existence?

Time is often seen as a linear progression—from past to present to future, eventually leading to an irreversible end. But if the universe follows cycles, then time itself may not be linear but circular, or even multi-dimensional.

1. Time as a Loop Instead of a Line

- In a cyclic universe, time does not start and stop—it resets and flows continuously, much like the seasons of nature.
- This challenges our perception of time as a one-way arrow leading to entropy and ultimate dissolution.
- Could time itself be a flexible construct that operates differently depending on cosmic cycles?

2. Does Time Exist Beyond the Universe?

- If universes are born, die, and are reborn, then time may only be meaningful within each individual cycle.
- This suggests that beyond the cosmic cycle, time may not exist in any conventional sense—it may emerge and dissolve with each universe.
- If so, then our understanding of time is not fundamental, but a property of the universe itself.

3. Reincarnation of the Cosmos—A Universal Principle?

- Many ancient traditions, including Hindu and Buddhist philosophy, describe cycles of rebirth—not just for individuals, but for entire worlds.
- Could this idea of cosmic reincarnation be a reflection of a deeper truth about the universe?
- If existence is cyclic and eternal, then nothing truly disappears—it only transforms.

This perspective radically changes how we think about existence, mortality, and the nature of reality itself.

How Does This Model Challenge the Singularity Concept and Redefine Space-Time?

The biggest challenge in modern cosmology is what happens at the extremes of the universe—both at its birth (the Big Bang) and at its end (black hole singularities).

1. The Death of the Singularity?

- The idea of a true singularity—a point of infinite density where physics breaks down—has always been problematic.
- If the universe cycles through expansion and contraction, then singularities may not exist at all.
- Instead of reaching infinite density, matter and energy may enter an ultra-compressed but finite state before rebounding.

2. A New Model of Space-Time as Elastic and Self-Regulating

- A cyclic universe suggests that space-time is not rigid but dynamic—expanding, contracting, and evolving across cosmic epochs.
- Instead of a space-time fabric that leads to dead ends, it may behave like a living system, constantly reorganizing itself.
- This could lead to new models of gravity, quantum mechanics, and even an alternative to the singularity problem in black holes.

3. Does Space-Time Remember Past Universes?

- If the universe has cycled through previous expansions and contractions, is there any memory of past cosmic cycles?
- Some physicists propose that traces of earlier universes could be encoded in the cosmic microwave background (CMB).
- Could understanding these remnants help us unlock the secrets of the past—and even predict the future of cosmic evolution?

If space-time is not just a passive structure but an active participant in cosmic cycles, then the very fabric of reality may be far more complex than we ever imagined.

Could Humanity's Understanding of Cosmic Cycles Lead to New Ways of Harnessing Energy, Space Travel, or Even Interacting with Black Holes?

If the universe naturally moves through cycles of creation and dissolution, then mastering its energy dynamics could unlock revolutionary new possibilities.

1. Tapping into Cosmic Energy Flows

- If black holes are not just destructive entities but cosmic energy regulators, could they be harnessed?
- Some theories suggest that black holes and the quantum vacuum contain vast amounts of untapped energy.
- Could understanding cosmic cycles help us develop technologies that use space-time energy, much like ancient civilizations described Prana as the fundamental force of existence?

2. Space Travel and Cosmic Gateways

- If black holes are not dead ends but portals to new cosmic phases, could we one day use them for interstellar or even interdimensional travel?
- The idea of wormholes (Einstein-Rosen bridges) suggests that space-time may have shortcuts linking different parts of the universe.
- Could mastering space-time's cyclical nature allow us to navigate cosmic structures beyond what we currently understand?

3. The Search for Past and Future Universes

- If the universe has cycled before, could advanced technology allow us to peer into past cosmic epochs?
- Some physicists speculate that gravitational waves and cosmic background radiation could contain signatures from previous universes.
- Could we one day develop ways to extract knowledge from past cosmic cycles, learning from the universe's infinite past?

If humanity can align itself with the natural rhythms of the universe, then we may unlock discoveries that transform our understanding of energy, time, and space travel.

Conclusion: The Future of Human Understanding in a Cyclic Universe

If the universe is part of an eternal cycle, then the way we think about time, energy, and space must evolve.

- Time may not be linear but an emergent property of the universe, appearing and dissolving with each cosmic cycle.
- Singularities may not exist, and instead, space-time could be elastic, preventing infinite collapse.
- Black holes may not be endpoints but key players in cosmic rebirth, preserving information and energy across cycles.
- Understanding cosmic cycles could lead to breakthroughs in energy, space travel, and even new ways of interacting with the fundamental forces of nature.

This raises some of the biggest questions ever asked:

- Is our current universe just one in an infinite series?
- Can we one day unlock knowledge from past cosmic cycles?
- Does existence itself follow a structured, self-repeating

process, where even consciousness may play a role in shaping reality?

If the universe is indeed an eternal, breathing entity, then our exploration of space and time has only just begun.

10: The Future of Cosmic Science – Merging Ancient Insights with Modern Exploration

In the previous chapter, we explored how the universe may not be a one-time event but a self-sustaining cycle of expansion and contraction, with black holes possibly acting as cosmic bridges between cycles. This view challenges the traditional perspective that space-time is a passive stage for cosmic evolution, instead suggesting that the universe operates through structured energy flows that regulate creation and dissolution. Yet, these ideas are not entirely new. Ancient civilizations—especially Vedic thinkers—conceived of the cosmos as a breathing entity, bound by rhythmic cycles of formation, transformation, and renewal. They described universal energy (Prana), cosmic cycles (Yugas and Kalpas), and the idea of dissolution (Pralaya), concepts that seem strikingly relevant to modern astrophysics.

But why has mainstream science largely dismissed these perspectives? Is it possible that ancient knowledge and modern physics are simply two ways of describing the same reality? If so, could integrating these insights open new frontiers in understanding cosmic evolution, black holes, space-time structures, and even the fundamental nature of energy?

As we stand at the threshold of a new era in astrophysics, the merger of ancient wisdom and modern exploration may offer a more complete understanding of how the universe functions—not as a chaotic, random system, but as a structured, self-regulating entity. This chapter explores why science should reconsider ancient perspectives, how structured energy models challenge existing theories, and how uniting physics and metaphysics could lead to revolutionary discoveries about the cosmos.

10.1 The Overlap Between Ancient Cosmology and Modern Physics

For centuries, science and spirituality have been seen as separate domains—one rooted in empirical observation, the other in mystical intuition. Yet, as modern physics delves deeper into the fundamental nature of space, time, and energy, it is uncovering patterns and principles that echo ancient cosmological insights. The striking parallels between Hindu, Buddhist, and Greek cosmologies and modern astrophysical theories suggest that ancient civilizations may have intuitively grasped the structured flow of energy in the cosmos—something science is only beginning to quantify.

Did these early thinkers, without telescopes or mathematical models, conceptualize space-time distortions, cosmic cycles, and even dark energy in symbolic terms? If so, could their insights help modern physicists reshape our understanding of the universe?

This section explores how ancient civilizations envisioned cosmic structures, the role of energy in creation and dissolution, and whether their intuitive understanding aligns with modern theories of space-time and energy dynamics.

How Early Civilizations Conceptualized the Cosmos in Ways That Mirror Today's Theories

Across ancient traditions, the universe was not seen as a static or random entity, but as an interconnected, evolving system. Many cultures described the cosmos as a living, breathing structure, bound by fundamental principles that governed its cycles. These ideas bear uncanny similarities to modern physics.

1. Hindu Cosmology: The Universe as a Cyclic Entity

- The Vedic concept of Yugas and Kalpas describes vast cycles of cosmic expansion and contraction, much like modern cyclic universe models.

- The idea of Pralaya (cosmic dissolution) aligns with theories of the Big Crunch or the oscillating universe.
- The notion of Brahman, an infinite, unchanging reality, resembles the idea of a quantum vacuum—the underlying energy field from which all matter arises.

2. Buddhist Cosmology: The Concept of Space-Time as an Illusion

- Buddhist teachings describe the impermanence of all things, including the cosmos itself, much like entropy-driven physical processes.
- The idea of dependent origination (Pratītyasamutpāda) suggests that nothing exists independently but arises due to interconnected causes—mirroring how quantum fields interact to create matter.
- Some interpretations of Buddhist philosophy even describe multiple universes coexisting, resembling modern multiverse theories.

3. Greek and Western Traditions: The Geometrical Universe

- Pythagorean and Platonic cosmology viewed the cosmos as an ordered, mathematical structure, much like today's emphasis on symmetry in physics.
- Heraclitus' concept of eternal change aligns with the idea that the universe undergoes continuous transformation rather than existing in a fixed state.
- The Stoic belief in a cyclic universe, destroyed and reborn through a cosmic fire (ekpyrosis), resembles the concept of a cyclic Big Bounce.

These ancient perspectives suggest that cosmic structures were not only philosophically understood but symbolically represented in ways that science is now beginning to recognize.

Why Hindu, Buddhist, and Greek Cosmological Models Describe Structured Energy Flows

One of the most striking similarities between ancient and modern cosmology is the concept of structured energy governing the universe. While modern physics describes this in terms of quantum fields, gravitational waves, and dark energy, ancient traditions spoke of Prana, Qi, or the cosmic balance of opposites.

1. The Hindu Concept of Prana and Universal Energy Flow

- In Hindu thought, Prana is not just life energy but a universal force that flows through all existence.
- This resembles modern theories of space-time energy fields, such as the quantum vacuum and dark energy.
- Just as Prana is said to maintain balance in the cosmos, modern physics suggests dark energy regulates cosmic expansion.

2. Buddhist Emptiness and the Vacuum Energy of Space

- Buddhist philosophy describes Śūnyatā (emptiness) as the fundamental state of reality, from which all things emerge and dissolve.
- This aligns with the concept of the quantum vacuum—an invisible field of fluctuating energy that gives rise to particles and forces.
- Both perspectives suggest that what appears as "nothingness" is actually a dynamic field full of potential energy.

3. Greek Cosmic Harmony and the Laws of Physics

- Pythagoras and Plato saw the universe as governed by mathematical harmony, which resonates with modern physicists' search for unified field equations.
- The belief that the cosmos follows predictable cycles of order and disorder aligns with the second law of thermodynamics.
- This structured approach to cosmic energy anticipates modern ideas of conserved quantities, equilibrium, and space-time symmetries.

If these civilizations intuited the existence of a structured, self-regulating cosmic energy system, could this knowledge help modern physics refine its models of dark energy, gravitational balance, and the nature of space-time?

Did Ancient Seers Intuitively Understand Space-Time Distortions, Cosmic Cycles, and Dark Energy?

While ancient civilizations lacked telescopes, satellites, and particle accelerators, their intuitive understanding of cosmic forces may have been encoded in their philosophies and spiritual teachings. Could they have perceived the effects of cosmic energy and gravitational distortions in ways that science is only now beginning to quantify?

1. The Concept of Time Dilation in Ancient Thought

- Hindu texts describe the relativity of time—how time flows differently in different cosmic states.
- The story of King Kakudmi meeting Brahma and realizing thousands of years had passed on Earth mirrors Einstein's theory of time dilation.
- The idea that time slows down in the presence of massive celestial bodies aligns with how black holes distort space-time.

2. Dark Energy and the Expansion of the Universe

- Ancient Hindu and Buddhist traditions describe the universe as constantly expanding and contracting, governed by an unseen force.
- Modern physics now recognizes that dark energy drives cosmic expansion, a concept that may have been metaphorically understood as an all-pervading cosmic force.
- If these traditions viewed energy as the fundamental driver of cosmic cycles, were they intuitively describing something akin to dark energy's effect on the cosmos?

3. The Vedic Concept of Multiverses and Quantum Reality

- Some Hindu texts describe multiple universes coexisting in different states of reality.
- This aligns with the many-worlds interpretation of quantum mechanics, which proposes that multiple versions of reality exist simultaneously.
- The idea that each cycle of creation spawns a new universe resembles modern theories of multiverse formation through cosmic inflation.

While these parallels do not serve as direct scientific evidence, they challenge the assumption that ancient knowledge was purely mythological. Instead, they suggest that ancient cosmology and modern physics are describing the same fundamental truths through different languages.

Conclusion: A Lost Scientific Perspective Waiting to Be Reclaimed?

As modern physics advances, it is beginning to echo the intuitive knowledge embedded in ancient cosmologies.

- The Hindu and Buddhist understanding of cyclic creation

aligns with modern cyclic universe theories.
- The concept of Prana as a structured energy flow resembles modern discussions of dark energy and vacuum fluctuations.
- Ancient descriptions of time distortion and cosmic expansion bear striking similarities to relativistic space-time models.

This raises an important question: Should science revisit ancient perspectives—not as myth, but as intuitive models that anticipated modern discoveries? If so, could integrating these insights lead to breakthroughs in space-time physics, energy balance, and cosmic evolution?

In the next section, we explore why the scientific community has largely ignored these ideas—and why it might be time to reconsider them.

10.2 Why the Scientific World Should Not Ignore Ancient Perspectives

For much of modern history, science and metaphysics have been treated as separate domains—one based on empirical observation and experimentation, the other rooted in intuition and philosophy. This divide between scientific materialism and ancient wisdom has led to the systematic dismissal of metaphysical concepts as mere mythology. However, as physics delves deeper into the nature of space, time, energy, and consciousness, it is increasingly aligning with ideas that ancient civilizations explored through intuition, meditation, and philosophical inquiry.

But why were these perspectives ignored in the first place? Did Western scientific thought, shaped by centuries of mechanistic reasoning, overlook valuable insights simply because they came from a non-empirical tradition? And most importantly, can modern science benefit from re-examining these ancient perspectives, not as myths, but as alternative models for understanding the cosmos?

This section explores the scientific bias against metaphysical knowledge, how quantum mechanics is forcing a paradigm shift, and whether integrating ancient insights could accelerate breakthroughs in cosmology.

The Western Scientific Bias – Why Metaphysical Concepts Were Dismissed as Mythology

Since the rise of modern science during the Age of Enlightenment, Western thought has been dominated by a mechanistic worldview—a perspective that sees the universe as a giant machine, operating solely through physical laws, without any inherent consciousness or deeper structure. This approach, while incredibly successful in advancing technology, has often rejected knowledge that does not fit within the rigid framework of empirical observation.

1. The Separation of Science and Spirituality

- The Scientific Revolution of the 16th and 17th centuries, led by figures like Newton and Galileo, emphasized reductionism—breaking the universe down into mechanical parts to be studied independently.
- This led to the separation of science from metaphysical traditions, which were often associated with religious or mystical ideas.
- Ancient civilizations, which viewed the universe as a dynamic, interconnected system governed by cycles and energy flows, were dismissed as having no real scientific basis.

2. Colonialism and the Rejection of Eastern Philosophies

- During the colonial era, Western scholars often dismissed Eastern knowledge systems as pre-scientific or superstitious.
- Despite profound mathematical, astronomical, and cosmological advancements in India, China, and the Middle

East, their contributions were largely ignored or appropriated without acknowledgment.
- The Vedas, Upanishads, and Buddhist Sutras contained deep cosmological insights, yet they were classified as religious texts rather than philosophical explorations of reality.

3. Materialism and the Neglect of Consciousness and Energy Fields

- Classical physics focused entirely on matter and forces, ignoring the possibility that energy flows, consciousness, or unseen forces (such as dark energy) might play a role in the cosmos.
- Concepts like Prana (life energy), Chi, or Akasha (the etheric field) were dismissed as mystical ideas rather than scientific hypotheses.
- This rigid materialist approach prevented science from exploring alternative models that could have led to earlier discoveries about quantum fields, dark energy, or the non-local nature of the universe.

However, this mindset is now being challenged by modern discoveries, particularly in quantum mechanics, which is revealing that reality is far more interconnected and fluid than classical physics once assumed.

How Quantum Mechanics Is Bringing Science Closer to Ancient Philosophies

The emergence of quantum physics in the 20th century shattered the rigid, deterministic framework of classical physics. Suddenly, the universe was no longer predictable and mechanistic—it was probabilistic, non-local, and fundamentally uncertain. Many of these discoveries bear striking similarities to ancient metaphysical concepts.

1. The Observer Effect and Consciousness

- In quantum mechanics, the act of observation affects reality—a concept that challenges the idea of an objective, independent universe.
- This aligns with the Vedantic and Buddhist view that consciousness plays a role in shaping reality.
- Could this mean that the ancient idea of the mind influencing the cosmos was not superstition, but an intuitive understanding of quantum mechanics?

2. Quantum Entanglement and Non-Locality

- Particles can be instantaneously connected across vast distances, defying classical ideas of space-time.
- Ancient Hindu and Buddhist texts speak of a universal interconnectedness, where all things arise from a single field of existence.
- Could quantum entanglement be scientific proof of the deep interconnectivity described in ancient cosmology?

3. The Quantum Vacuum and the Concept of Akasha

- Quantum physics suggests that "empty space" is not empty—it is a seething field of energy, full of fluctuations and potentiality.
- The Vedic concept of Akasha describes an all-pervasive cosmic field that contains the fundamental energy of creation.
- Could modern physics be rediscovering what ancient thinkers intuited thousands of years ago?

4. Multiple Realities and the Nature of Time

- The Many-Worlds Interpretation of quantum mechanics suggests that multiple parallel realities may exist.

- Hindu cosmology describes multiple dimensions, cycles of reality, and time as non-linear.
- Could it be that the ancients had a deep understanding of these cosmic principles, but expressed them in symbolic rather than mathematical language?

These parallels suggest that ancient knowledge may not be outdated mythology, but a different way of understanding the same fundamental truths that physics is now exploring.

Could Integrating Ancient Insights Accelerate Breakthroughs in Cosmology?

If ancient perspectives align so closely with modern discoveries, should scientists actively study and integrate these insights into their research? Could looking at the universe through the lens of structured energy, interconnectedness, and cyclic transformation help us solve some of the biggest mysteries in astrophysics?

1. A New Approach to Dark Energy and Cosmic Expansion

- Dark energy is a mysterious force causing the universe to expand at an accelerating rate.
- Ancient traditions describe an all-pervading cosmic energy (Prana, Chi, Akasha) that influences matter and movement.
- Could these concepts offer a new way to interpret dark energy—not as a separate force, but as a fundamental aspect of space-time itself?

2. Redefining Space-Time as a Dynamic, Living System

- Instead of seeing space-time as a passive stage for cosmic evolution, ancient models suggest that it is an active participant, regulating cycles of energy.
- This could lead to new physics models that incorporate

structured energy flows instead of relying solely on gravity and entropy.

3. Unlocking New Models for Interstellar Travel

- If space-time is not just a rigid fabric but a fluctuating energy field, could future technologies harness this knowledge for advanced space travel?
- Ancient texts speak of portals, dimensional shifts, and cosmic pathways—ideas that, if studied scientifically, could inspire breakthroughs in propulsion physics.

4. The Role of Consciousness in Physics

- If consciousness plays a role in shaping reality, could this lead to new models of understanding quantum phenomena, space-time manipulation, or even the nature of existence itself?

By bridging the gap between ancient insights and modern physics, scientists may unlock revolutionary discoveries that could reshape our understanding of the cosmos.

Conclusion: A Paradigm Shift in Cosmic Science

The scientific world has long dismissed ancient perspectives as unscientific, yet modern discoveries are revealing that many of these ideas were deeply insightful.

- Quantum mechanics, space-time theories, and energy flow models are now aligning with ancient cosmology.
- Integrating metaphysical insights could lead to breakthroughs in astrophysics, dark energy research, and space exploration.
- The universe may not be a cold, mechanical system—but a dynamic, evolving, and interconnected reality.

If science and ancient wisdom are describing the same truths in different languages, then merging them may be the key to unlocking the deepest mysteries of existence.

10.3 Structured Energy Models: Challenging the Singularity Paradigm

For decades, the concept of singularities—points of infinite density where physical laws break down—has been one of the biggest paradoxes in astrophysics. Black holes, the Big Bang, and the potential Big Crunch are all thought to involve singularities, yet they create fundamental contradictions in physics. The very existence of singularities suggests that our understanding of space-time is incomplete.

But what if singularities are not real? What if instead of breaking down into an undefined state, space-time follows structured energy dynamics that regulate the formation, evolution, and transformation of cosmic structures?

Recent research in quantum gravity, dark energy, and black hole thermodynamics suggests that the universe does not collapse into infinite singularities, but instead operates through self-regulating energy structures. This aligns with ancient cosmological perspectives, which describe the cosmos as a dynamic, structured entity governed by cycles of creation and transformation rather than abrupt, absolute endings.

Could structured energy models provide the missing link between General Relativity and Quantum Mechanics? Could they help us move beyond singularities and develop a more complete theory of space-time?

This section explores why the singularity paradigm is problematic, how black holes and dark energy suggest an evolving space-time fabric, and whether structured energy could be the missing key to uniting gravity and quantum physics.

Why the Singularity Concept Is Problematic and May Be Replaced with Structured Energy Models

Singularities are often described as regions where gravity becomes infinite, space-time curves infinitely, and the known laws of physics cease to function. However, these ideas create serious contradictions in modern physics.

1. Infinite Density and the Breakdown of Physical Laws

- In General Relativity, singularities emerge because gravity becomes infinitely strong at a single point.
- But infinity is not a physical quantity—it is a mathematical abstraction that suggests we are missing key details about how energy behaves under extreme conditions.

2. Singularities Defy Quantum Mechanics

- Quantum mechanics describes reality at the smallest scales, but it cannot explain singularities.
- In a singularity, matter and energy collapse into an undefined state, contradicting quantum principles that require measurable, probabilistic behavior.
- This suggests that something else—perhaps structured energy dynamics—must regulate these extreme states.

3. The Cosmic Censorship Problem

- If singularities truly exist, why don't we see them everywhere?
- Physicist Roger Penrose's Cosmic Censorship Hypothesis suggests that nature somehow hides singularities behind event horizons (as in black holes), preventing them from directly influencing space-time.
- Could this mean that singularities do not actually exist, but are instead energy structures that behave in an ordered way?

These issues indicate that singularities are not fundamental truths, but signs that our current models are incomplete. This is where structured energy models offer an alternative approach.

How Black Holes, Dark Energy, and Cosmic Cycles Suggest an Evolving Space-Time Fabric

If singularities are not real, then how do black holes and cosmic expansion function? The answer may lie in structured energy flows that regulate how gravity, time, and matter interact at extreme scales.

1. Black Holes as Energy Regulators, Not Singularities

- Instead of being infinitely dense points, black holes may have structured energy cores that redistribute matter and energy.
- Some theories suggest that black holes do not trap energy forever, but instead cycle it through different phases—potentially even into new universes.
- This aligns with Hindu cosmology's idea of cosmic transformation rather than absolute endings.

2. Dark Energy as a Balancing Force in Cosmic Evolution

- Dark energy is accelerating the expansion of the universe, but we do not yet understand its true nature.
- Could it be part of a structured energy system that prevents singularities from forming by maintaining cosmic balance?
- Some physicists suggest that dark energy is not a separate force, but a property of space-time itself—echoing ancient ideas of an all-pervading cosmic energy (Prana, Akasha).

3. Cyclic Universes and the Death of the Singularity

- The idea of a cyclic universe (Big Bounce) suggests that space-time never fully collapses into a singularity.
- Instead, at the moment of maximum contraction, quantum

effects create a new expansion phase, preventing an infinite breakdown.

- This would mean that the universe is self-regulating, structured, and infinite—not a one-time event leading to an undefined singularity.

These insights suggest that space-time does not break down into singularities, but instead follows structured energy laws that allow for continuous cosmic evolution.

Could Structured Energy Be the Missing Link Between Relativity and Quantum Mechanics?

One of the biggest challenges in modern physics is unifying General Relativity (which describes gravity on large scales) with Quantum Mechanics (which describes the microscopic world). Singularities highlight the disconnect between these theories—but structured energy models may offer a solution.

1. A New Understanding of Gravity as Energy Flow

- Instead of treating gravity as a force acting between objects, structured energy models suggest that gravity emerges from the flow of energy within space-time.
- This aligns with the idea that black holes, cosmic expansion, and quantum fields are all interconnected through energy transfer rather than fixed equations.

2. Quantum Gravity Without Singularities

- Some theories, such as Loop Quantum Gravity (LQG), propose that space-time is made of discrete energy loops rather than a continuous fabric.
- If true, this would mean that singularities cannot form because space-time has a minimum unit of energy structure.

- This supports the idea that black holes and cosmic evolution are governed by structured energy rather than infinite collapse.

3. The Role of Consciousness in Energy Dynamics

- Some physicists suggest that consciousness itself may be linked to energy flows in space-time.
- Ancient traditions describe mind and energy as interconnected (Prana, Qi, or Akasha interacting with thought).
- If space-time follows structured energy principles, could human consciousness be another form of structured energy?

By replacing singularities with structured energy dynamics, we may finally bridge the gap between quantum mechanics and gravity—unlocking a deeper understanding of space-time itself.

Conclusion: Toward a New Cosmic Paradigm

The singularity paradigm has long been a barrier to understanding black holes, cosmic evolution, and the fundamental nature of space-time. But emerging evidence suggests that the universe operates through structured energy flows rather than infinite collapse.

- Black holes may not contain singularities, but instead act as structured energy hubs that recycle matter and information.
- Dark energy could be part of a larger energy system regulating cosmic balance rather than an unknown force.
- Cyclic universe models suggest that the cosmos self-regulates rather than ending in undefined singularities.

By adopting a structured energy approach, scientists may be able to unite relativity and quantum mechanics, refine our understanding of space-time, and unlock a deeper, more complete vision of the cosmos.

Could this perspective also open the door to new forms of energy harnessing, space travel, or even insights into the nature of consciousness?

In the next section, we explore how a space-time energy model could lead to revolutionary advancements in cosmic physics, technology, and even our understanding of reality itself.

10.4 The Future of Space-Time Theory: Cosmic Energy as an Active Force

For centuries, space-time was viewed as a passive stage upon which cosmic events unfolded—a vast, empty void through which matter and energy moved. However, modern physics is beginning to reveal that space-time itself is dynamic, responsive, and structured. This shift challenges the mechanistic worldview that has dominated science since Newton and points toward a more energy-driven, interactive cosmos.

Ancient traditions have long described the universe as permeated by a universal life force—Prana, Qi, or Akasha—governing movement, balance, and transformation. Could this ancient idea of an all-pervasive cosmic energy align with emerging theories of quantum fields, dark energy, and gravitational waves?

If space-time is not just a neutral backdrop but an active participant in cosmic evolution, what new physics could emerge? Could this understanding lead to technological advancements, deeper insights into consciousness, and a radical rethinking of the fundamental nature of reality?

This section explores how the concept of Prana aligns with modern physics, how the universe is shifting from a mechanistic model to an energy-driven system, and what new breakthroughs could emerge from studying space-time as a self-regulating field.

How the Concept of Prana (Universal Energy) Aligns with Modern Physics

Across ancient traditions, Prana (Hindu), Qi (Chinese), and Ether (Greek) were understood as subtle forces that sustain and regulate the universe. Though dismissed for centuries as mystical beliefs, recent discoveries suggest that these ancient concepts may have an unexpected parallel in physics.

1. Prana and the Zero-Point Energy Field

- Quantum field theory proposes that even in a perfect vacuum, space is never truly empty—it is filled with energy fluctuations known as zero-point energy.
- This invisible, ever-present energy aligns with the idea of Prana as a fundamental, all-pervading force that underlies reality.
- Could what ancient sages intuited as Prana actually be the dynamic, quantum energy field that permeates space-time?

2. Prana as the Missing Link in Dark Energy

- Dark energy is responsible for the accelerating expansion of the universe, yet scientists still do not understand its true nature.
- Some theorists propose that dark energy is not a separate force but an intrinsic property of space-time, similar to how Prana is described as an inherent energy flow in the cosmos.
- Could understanding dark energy through the lens of Prana help redefine space-time physics?

3. Prana, Consciousness, and the Observer Effect

- In quantum mechanics, the observer effect suggests that conscious observation influences physical reality.
- Ancient traditions describe Prana as not just energy but also as a force linked to awareness and life itself.

- If consciousness plays a role in shaping quantum events, could Prana be an early interpretation of the quantum-consciousness connection?

By revisiting Prana through the lens of modern physics, we may uncover a more holistic framework for understanding space-time, energy, and the fundamental nature of the universe.

The Shift from a Mechanistic Universe to an Energy-Driven, Dynamic Cosmos

For most of modern history, the universe was viewed as a machine—governed by fixed laws, with energy acting as a passive quantity that simply moved through it. However, new discoveries are challenging this view, revealing that the universe may be fundamentally energetic, self-organizing, and responsive.

1. The Death of the Static Universe

- Classical physics imagined space as a passive void, with time flowing in one direction.
- Einstein's General Relativity changed this, showing that space-time itself bends and warps due to energy and gravity.
- Today, dark energy and quantum fields suggest that space-time is an active, ever-evolving medium rather than a rigid, empty expanse.

2. Matter Is Not Fundamental—Energy Is

- Einstein's famous equation $E = mc^2$ revealed that matter is just a condensed form of energy.
- The deeper we explore physics, the more it becomes clear that energy is the true foundation of existence, not physical matter.
- This aligns with ancient ideas where energy (Prana, Qi, Akasha) was seen as the primary force shaping reality.

3. Space-Time as a Self-Regulating Field

- Some modern physicists propose that space-time is not passive but dynamically self-regulating, much like an ecosystem.
- This idea resembles ancient cosmology, where energy was described as flowing in structured cycles, constantly balancing itself.
- If space-time is an intelligent system that adjusts based on cosmic energy flows, could this unlock new theories of gravity, dark matter, and quantum behavior?

This paradigm shift—from a mechanistic cosmos to an energy-driven, dynamic universe—marks one of the most profound transformations in science.

What New Models Could Emerge by Studying Space-Time as a Self-Regulating System?

If space-time is not just a neutral background but an active field of energy exchange, what new possibilities does this open? Could this lead to a new era of space travel, energy manipulation, and deeper insights into the cosmos?

1. A Unified Theory of Space-Time and Energy

- Current physics struggles to unify General Relativity (gravity) with Quantum Mechanics (subatomic energy).
- A structured energy model could provide a common framework, integrating both cosmic and quantum scales into a single system.
- This could redefine the very nature of forces, fields, and the interaction between energy and matter.

2. Harnessing the Energy of Space-Time

- If space-time itself contains structured energy flows, could future technology tap into these forces?

- Zero-point energy, quantum fluctuations, and gravitational waves could be key to unlocking new forms of propulsion, energy harvesting, and interstellar travel.
- Some speculate that ancient civilizations may have already experimented with such energy principles in lost sciences.

3. Rewriting the Laws of Motion and Gravity

- If space-time is not fixed but fluid, then gravity may be an emergent force rather than a fundamental one.
- Some new theories propose that gravity arises from the flow of quantum information—an idea that aligns with ancient views of energy balance in the universe.
- Could understanding space-time as an adaptive, energy-driven system allow us to manipulate gravity, travel faster than light, or even interact with black holes?

4. New Insights into Consciousness and Reality

- If energy and consciousness are linked, could space-time itself have a form of awareness?
- Some physicists suggest that consciousness arises from quantum processes, meaning reality itself may be a self-aware, self-regulating system.
- This would echo ancient ideas that the universe is not just physical, but intelligent—a living field of energy and information.

By studying space-time as a structured, self-regulating system, humanity may be on the verge of unlocking a new physics—one that integrates energy, gravity, and consciousness into a unified vision of reality.

Conclusion: A New Cosmic Vision Based on Energy and Interconnectivity

The idea that space-time is a passive void is rapidly being replaced by a vision of the universe as an active, energy-driven system.

- Prana, Qi, and Akasha may represent early understandings of quantum energy fields, dark energy, and space-time structure.
- The universe is shifting from a mechanistic view to an energy-based perspective, where space-time is self-organizing and dynamic.
- New models of physics could emerge by embracing space-time as an interactive energy system, potentially leading to revolutionary advancements in technology, propulsion, and our understanding of consciousness.

As we step into this new era of science, we may not only redefine physics but also reconnect with ancient knowledge—revealing that the greatest cosmic truths were always known, waiting to be rediscovered.

In the next section, we explore how integrating metaphysics and physics can lead to a more complete understanding of the universe—one where ancient wisdom and modern exploration unite to reveal the deepest mysteries of existence.

10.5 A New Way Forward: Uniting Metaphysics and Physics to Decode the Universe

The exploration of the cosmos has long been dominated by empirical science, mathematical models, and technological advancements. Yet, as physics pushes deeper into the nature of reality—exploring dark energy, quantum entanglement, and the fabric of space-time—it is beginning to encounter profound mysteries that align with ancient metaphysical concepts.

For centuries, metaphysical traditions such as Hindu cosmology, Buddhist philosophy, and Greek thought have described the universe as a self-regulating, interconnected system, governed by energy flows, consciousness, and cyclic time. While these ideas were once dismissed as myth or spiritual symbolism, modern science is now uncovering evidence that suggests these ancient perspectives may hold valuable insights into the structure of reality.

If space-time is a structured energy system, and if consciousness plays a role in shaping physical reality, could merging metaphysics with astrophysics unlock new discoveries? Could a holistic approach—combining ancient wisdom with cutting-edge technology—lead to a paradigm shift in our understanding of the universe?

This section explores why a more integrative approach is needed, how metaphysical concepts like energy balance and cyclic time can be incorporated into scientific models, and how combining ancient insights with modern advancements may reshape cosmic exploration.

The Need for a More Holistic Approach—Bridging Astrophysics with Ancient Knowledge

The divide between science and metaphysics has led to a fragmented understanding of reality, where certain aspects of existence (such as consciousness, time perception, and energy balance) are excluded from mainstream physics. However, recent breakthroughs suggest that a more holistic approach—one that unites astrophysics with ancient cosmology—could provide a deeper, more accurate understanding of the cosmos.

1. Beyond Reductionism: The Shift Toward a Systems Approach

- Classical physics has long followed a reductionist approach—breaking the universe down into particles, forces, and isolated laws.
- However, modern physics is shifting toward a systems

approach, recognizing that the universe functions as an interconnected whole.
- This aligns with ancient cosmological models, which viewed the cosmos as a dynamic, self-organizing system rather than a collection of separate components.

2. The Role of Energy Flow and Cosmic Equilibrium

- Hindu cosmology describes the universe as a balance between forces—creation, preservation, and dissolution.
- This concept closely mirrors modern discussions about entropy, dark energy, and the balance between gravitational forces and cosmic expansion.
- If ancient traditions understood energy regulation without modern mathematics, could integrating these insights lead to new discoveries about cosmic equilibrium?

3. Revisiting Ancient Models with Scientific Tools

- Many ancient descriptions of cosmic cycles, energy transformation, and the structure of reality resemble modern astrophysical theories.
- Instead of dismissing these ideas as mythology, what if they were early observational models that require reinterpretation through the lens of contemporary science?
- By scientifically analyzing ancient texts, symbols, and oral traditions, we may uncover lost knowledge that aligns with cutting-edge physics.

To move forward, we must bridge the gap between modern astrophysics and ancient insights, creating a unified approach that embraces the strengths of both perspectives.

Can Metaphysical Concepts Like Consciousness, Energy Balance, and Cyclic Time Be Integrated into Scientific Models?

While science has traditionally separated consciousness from physics, emerging research in quantum mechanics, neurophysics, and cosmology suggests that consciousness and physical reality may be more deeply connected than previously believed.

1. The Observer Effect and the Role of Consciousness

- Quantum experiments have demonstrated that observation influences reality, leading some scientists to propose that consciousness may play a fundamental role in shaping the physical world.
- Ancient traditions have long suggested that the universe is not purely mechanical, but shaped by awareness, perception, and energy flow.
- If consciousness is more than just a byproduct of brain activity, could it be an active force in the cosmos, influencing quantum fields and space-time?

2. Cyclic Time and the Nature of Cosmic Evolution

- Hindu and Buddhist cosmology describe time as cyclic, rather than linear—a perspective that aligns with cyclic universe models in physics.
- If the universe undergoes repeating cycles of expansion and contraction (Big Bang and Big Crunch, or Big Bounce theories), could these cycles be encoded in the fabric of space-time?
- Understanding time as an emergent property of space-time rather than a fixed quantity could redefine astrophysical models.

3. The Interplay Between Energy, Matter, and Information

- Many metaphysical traditions describe the universe as an interplay of energy, vibration, and structured information.
- Modern physics is beginning to recognize the role of information in fundamental physics, such as black hole entropy and holographic principles.
- Could energy, matter, and consciousness all be different expressions of a deeper underlying informational structure within the universe?

These ideas suggest that integrating metaphysical insights into scientific models could help answer some of the deepest unanswered questions in cosmology.

The Next Frontier: How Combining Ancient Insights with Modern Technology May Lead to a Paradigm Shift in Cosmic Exploration

As science advances, we are approaching a new frontier—one where the boundaries between physics, consciousness, and cosmic energy must be re-evaluated.

By combining ancient wisdom with cutting-edge technology, we may be able to unlock new ways of understanding the cosmos, exploring space, and even harnessing fundamental forces in ways never imagined before.

1. Revolutionizing Space Travel with Energy-Based Propulsion

- If space-time is structured and dynamic, could we develop propulsion systems that interact with its energy flows rather than relying solely on fuel-based technology?
- Some advanced propulsion concepts, such as negative energy warp drives and gravitational field manipulation, suggest that future space travel may depend on understanding energy

structures in space-time.
- If ancient civilizations had knowledge of non-conventional energy forces, could we rediscover lost principles that allow for advanced propulsion?

2. Understanding and Harnessing Space-Time Energy

- Modern physics is already exploring zero-point energy and vacuum fluctuations as potential energy sources.
- Ancient traditions describe energy fields (Prana, Akasha) that sustain existence—could these ideas guide us in extracting usable energy from space-time itself?
- The potential of harnessing cosmic energy flows could revolutionize our ability to sustain civilizations beyond Earth.

3. Consciousness, Reality, and the Future of Human Understanding

- If consciousness interacts with quantum fields, what does this mean for future explorations of space-time, alternate dimensions, and even extraterrestrial intelligence?
- Could ancient meditation practices, once seen as purely spiritual, actually be techniques for tuning into deeper energy fields that science is only beginning to detect?
- If the universe is a structured, conscious system, how might human perception itself evolve in response to a greater cosmic awareness?

By integrating technology, physics, and metaphysical wisdom, we may enter an era where our understanding of the universe is no longer limited by outdated paradigms.

Conclusion: A New Age of Cosmic Discovery

Science and metaphysics have long been treated as separate paths, but the time has come to merge them into a unified approach to understanding the universe.

- Astrophysics must embrace structured energy models and cyclic cosmology, moving beyond the singularity paradigm.
- Consciousness, energy balance, and cosmic cycles must be explored scientifically, as they may be fundamental components of reality.
- By combining ancient insights with cutting-edge technology, we may uncover the next great leap in human understanding, unlocking cosmic secrets that could redefine existence itself.

In the coming decades, the intersection of physics, metaphysics, and cosmic exploration may lead to discoveries that bridge the gap between science and spirituality, uniting humanity's ancient wisdom with its future potential.

Conclusion: The Eternal Dance of the Cosmos

As we have traversed the vast intellectual and philosophical landscape of cosmic science, it becomes clear that the universe is not merely a collection of celestial objects governed by blind forces. Instead, it is a dynamic, evolving entity engaged in a perpetual dance of creation, preservation, and dissolution. This eternal cycle—mirrored in both ancient cosmology and modern astrophysics—reveals a deeper order, a hidden rhythm underlying all of existence.

The journey through this book has illuminated the striking parallels between Vedic cosmology and contemporary physics. From the concept of Hiranyagarbha as a primordial cosmic seed to the scientific model of the Big Bang, from the cyclic nature of Kalpas and Pralaya to theories of the Big Crunch and Big Bounce, we see that ancient wisdom often anticipated the discoveries of modern science. The Chavan's Limits, introduced in this book, challenge the notion of singularities by proposing structured energy equilibrium within black holes, redefining our understanding of cosmic balance.

Yet, the mysteries of the universe remain vast and largely unexplored. Black holes, once seen as endpoints of matter and energy, now emerge as potential bridges to new realms of physics—perhaps even to new universes. Dark energy, the enigmatic force driving cosmic expansion, may not be an unknown repulsive force but rather a structured energy redistribution mechanism akin to the ancient concept of Prana. The laws of physics are no longer seen as rigid, deterministic frameworks; instead, they appear to be expressions of deeper, self-organizing principles that govern cosmic evolution.

Dissolution is Not the End—It is a Gateway to New Beginnings

One of the most profound insights from both modern cosmology and ancient philosophy is that dissolution does not signify finality but transformation. Pralaya, much like the end of a cosmic cycle, is not an absolute erasure of existence but a transition into another phase of reality. Similarly, black holes do not merely destroy matter but may seed the birth of new universes. The Big Crunch, if it occurs, may not signal the death of the cosmos but its rebirth into a new expansion cycle, akin to Brahma's Day and Night.

If we view the universe through this lens, then we must also reconsider our place within it. Humanity has long sought to define its role in the cosmos through mythology, religion, philosophy, and science. But perhaps our greatest responsibility is not merely to observe the universe, but to align ourselves with its rhythms—to understand and embrace its cycles rather than resist them.

The Hidden Wisdom in Pralaya and Black Hole Physics

For centuries, the concept of Pralaya has been misinterpreted as mere destruction. However, when examined through the lens of physics, it appears to be a profound insight into the cosmic process of renewal. Similarly, black holes—once thought to be the ultimate voids—are now being reconsidered as energy regulators, recycling cosmic matter and potentially giving rise to new universes.

This shift in understanding represents an important moment in both science and philosophy. What was once feared as annihilation may now be seen as the foundation of new creation. Whether in the death of stars, the collapse of galaxies, or the rebirth of cosmic structures, we see the same recurring principle: Nothing is lost; everything is transformed.

Where Do We Go from Here? The Need for a Paradigm Shift in Cosmic Science

The insights from this book lead us to an essential realization: The time has come for a paradigm shift in cosmic science. The rigid separation between ancient philosophy and modern physics has limited our understanding of the universe. It is no longer sufficient to view these perspectives as oppositional; instead, they must be seen as complementary.

Science, in its pursuit of objectivity, often dismisses the intuitive and philosophical insights of ancient traditions. But history has shown us that some of the most profound discoveries arise when ideas from different disciplines intersect. If we integrate structured energy models, reconsider the role of black holes, and acknowledge the cyclic nature of the universe, we may uncover new fundamental principles that reshape our understanding of space-time.

Furthermore, the concept of consciousness as an active force in cosmic evolution—once confined to metaphysical discussions—may hold key insights into the nature of reality. Just as quantum physics has revealed that observation affects the behavior of particles, could the act of perception itself be woven into the fabric of cosmic rebirth? Could the very existence of life and intelligence be integral to the universe's self-regulating system?

Final Thoughts: Embracing the Cosmic Perspective

As we conclude this exploration, one truth emerges with clarity: The cosmos is not a static entity but a living, evolving system. Whether described through the equations of physics or the verses of ancient scriptures, its fundamental nature remains the same—an infinite cycle of transformation.

We, as conscious beings, are not separate from this dance of the cosmos. We are active participants in its unfolding reality. By understanding and embracing the universe's cycles, we align ourselves with the deeper forces that govern existence. This perspective is not merely an intellectual exercise—it is a call to expand our vision, to embrace the grand interconnectedness of all things, and to seek knowledge beyond the limitations of a single discipline.

The eternal dance of the cosmos continues. Whether we watch from the periphery or choose to step into its rhythm is a decision that shapes not only our understanding of the universe but also our place within it.

Acknowledgments

Writing *The Cosmic Breath: Black Holes, Pralaya, and the Energy Cycles of the Universe* has been an extraordinary journey, one that would not have been possible without the support, insights, and encouragement of many individuals.

First and foremost, I extend my deepest gratitude to my mentors, colleagues, and fellow researchers who have inspired and challenged my thinking in both scientific and philosophical realms. Your discussions and guidance have helped shape the ideas explored in this book.

A heartfelt thank you to my friends and family, whose unwavering support has given me the space and motivation to pursue this ambitious project. Your belief in my vision has been invaluable.

I also acknowledge the ancient wisdom of Vedic texts, Upanishads, and the profound teachings of Hindu cosmology, which have provided the foundation for exploring cosmic cycles. Their deep insights, when placed alongside modern astrophysics, reveal a universe far more structured and interconnected than we often imagine.

Finally, I express my appreciation to all those who seek knowledge and truth—scientists, philosophers, and readers like you—who continue to question, explore, and push the boundaries of our understanding. It is this collective pursuit of wisdom that will lead us to deeper truths about the cosmos and our place within it.

This book is dedicated to the eternal seekers—those who look at the stars and wonder, not just about what is, but about what could be.

References

The following sources have informed and inspired the exploration of cosmic cycles, black hole physics, and the parallels between ancient wisdom and modern astrophysics in *The Cosmic Breath: Black Holes, Pralaya, and the Energy Cycles of the Universe*. These works include **scientific research, historical texts, and key references** that provide further insights into the themes discussed in this book.

Author's Research and Publications

1. Chavan, S. (2024). *Beyond Singularity: Black Holes as Structured Energy Equilibrium*. Zenodo. https://doi.org/10.5281/zenodo.14793162
2. Chavan, S. (2024). *Pralaya and the Cosmic Breath: The Cyclic Nature of the Universe*. Zenodo. https://doi.org/10.5281/zenodo.14791053
3. Chavan, S. (2024). *Dark Energy and the Unseen Forces Governing the Cosmos*. Zenodo. https://doi.org/10.5281/zenodo.14788539
4. Chavan, S. (2024). *The Hidden Order: Energy Flow in Black Holes and Universal Expansion*. Zenodo. https://doi.org/10.5281/zenodo.14782900

Books and Academic References *(If applicable, add any additional citations)*

- Hawking, S. (1988). *A Brief History of Time: From the Big Bang to Black Holes.* Bantam Books.
- Penrose, R. (2004). *The Road to Reality: A Complete Guide to*

the Laws of the Universe. Alfred A. Knopf.

- Carroll, S. (2019). *Something Deeply Hidden: Quantum Worlds and the Emergence of Spacetime.* Dutton.
- Kaku, M. (2014). *The Future of the Mind: The Scientific Quest to Understand, Enhance, and Empower the Mind.* Doubleday.
- Mukunda, N., & V. Srinivasan. (2017). *Cosmology in the Indian Philosophical Tradition: The Cycles of Time and the Universe.* Indian Academy of Sciences.

Ancient Texts and Translations

- The *Rigveda* (Nasadiya Sukta – Hymn of Creation)
- The *Bhagavad Gita*, translated by Eknath Easwaran
- The *Upanishads*, translated by Swami Sivananda
- The *Surya Siddhanta* (ancient Hindu astronomical treatise)
- The *Brahmanda Purana* (cosmological descriptions in Hindu philosophy)

Scientific Papers and Research *(Optional – add specific papers cited in the book)*

- Einstein, A. (1915). *The Field Equations of Gravitation.*
- Bekenstein, J. D. (1973). *Black Holes and Entropy. Physical Review D.*
- Hawking, S. (1974). *Black Hole Explosions? Nature,* 248(5443), 30-31.
- Guth, A. (1981). *Inflationary Universe: A Possible Solution to the Horizon and Flatness Problems. Physical Review D.*

This collection of references provides a **comprehensive foundation** for readers interested in **further exploring cosmic cycles, black hole** physics, and the convergence of astrophysics with ancient cosmology.

physics, and the convergence of astrophysics with ancient cosmology.

About the Author

Sandeep Chavan is an engineer, researcher, educator, and independent thinker who has spent years exploring the intersections between astrophysics, ancient cosmology, and structured energy models of the universe. With a background in mechanical engineering and industrial research, he has a deep passion for uncovering scientific parallels in historical wisdom and integrating philosophical insights with modern physics.

His journey into cosmic science and universal theories began with a curiosity about the cyclical nature of time, energy balance, and the transformative role of black holes in the universe. Inspired by both cutting-edge astrophysics and Vedic cosmology, he has dedicated his research to challenging traditional paradigms, particularly the singularity model, and advocating for structured energy equilibrium as the foundation of cosmic evolution.

Sandeep has authored multiple works on black hole physics, Pralaya (universal dissolution), dark energy, and cyclic universe theories, with his research published on platforms such as Zenodo. His writings offer a unique blend of scientific depth and philosophical inquiry, making complex ideas accessible to both academics and general readers.

In *The Cosmic Breath: Black Holes, Pralaya, and the Energy Cycles of the Universe*, he seeks to bridge the gap between modern astrophysics and ancient knowledge, encouraging a paradigm shift in how we understand space-time, cosmic cycles, and the ultimate fate of the universe.

When he is not writing or researching, Sandeep actively mentors students and professionals in engineering, physics, and universal theories, fostering a deeper understanding of science beyond conventional academic boundaries.

Connect with the Author:

Readers can explore more of Sandeep Chavan's work, research, and insights at:

Website: www.sandeepjchavan.com[1]

Research on Zenodo: Zenodo Profile[2]

Published Books: Available on Amazon, Google Books, Apple Books, and other platforms.

1. https://www.sandeepjchavan.com

2. https://zenodo.org/

Final Thoughts

As we stand on the precipice of a new era in cosmic exploration, one thing becomes increasingly clear—the universe is far more interconnected, dynamic, and structured than we ever imagined. The lines between science and spirituality, physics and philosophy, ancient wisdom and modern discovery are no longer as rigid as they once seemed. Instead, they are converging, offering us a more holistic understanding of reality.

This book has explored the parallels between Vedic cosmology and astrophysics, revealing that the ancient concept of Pralaya mirrors black hole physics, that cyclic universe theories align with the Hindu vision of cosmic rebirth, and that structured energy models may redefine our understanding of space-time. These ideas, once thought to be separate, are now part of an evolving dialogue between metaphysics and modern science.

Yet, many questions remain unanswered.

- What truly happens beyond the event horizon of a black hole?
- Does the universe undergo infinite cycles of expansion and contraction, or does it evolve in unforeseen ways?
- Is consciousness an active force in cosmic evolution, or is it merely an emergent property of matter?
- Can space-time itself be manipulated, harnessed, or understood as an intelligent system?

The future of cosmic science will not be limited to mathematical equations and telescopic observations alone—it will require a shift in perspective, an openness to integrating ancient insights, and a willingness to challenge the very foundations of physics as we know it.

As humanity ventures deeper into space, building new technologies, uncovering hidden cosmic forces, and pushing the limits of knowledge, we must also look inward—toward our shared history, our ancient texts, and the timeless wisdom that may hold clues to the universe's deepest mysteries.

The quest to understand the cosmic breath—the eternal rhythm of creation, dissolution, and rebirth—is far from over. In many ways, we are just beginning to grasp the profound elegance of the universe. And as we move forward, the answers we seek may not just lie in distant galaxies, but in the wisdom we have carried with us all along.

The universe breathes. Let us listen. Let us explore. Let us evolve.

Disclaimer

The information presented in *The Cosmic Breath: Black Holes, Pralaya, and the Energy Cycles of the Universe* is intended for educational and thought-provoking purposes. This book explores the intersections between modern astrophysics, ancient cosmology, and structured energy models, offering theoretical interpretations and philosophical insights that do not claim to represent definitive scientific conclusions. While references to scientific research, historical texts, and Vedic cosmology are included, the content should not be interpreted as peer-reviewed scientific findings, religious doctrine, or a substitute for formal academic research. The perspectives shared in this book are intended to encourage critical thinking, interdisciplinary exploration, and open-ended discussion on the nature of the universe.

Readers are encouraged to engage with primary scientific literature, ancient texts, and academic sources for a deeper understanding of the topics covered. The author assumes no responsibility for the interpretation or application of the ideas presented in this book.

Any mention of historical philosophies, religious texts, or scientific theories is purely for comparative analysis and does not imply endorsement or absolute validation of any particular worldview. The book does not aim to dispute established scientific principles but rather to expand perspectives by bridging diverse domains of knowledge.

The views expressed are those of the author and do not represent the opinions of any institution, organization, or governing body.

By reading this book, the reader acknowledges that the content is exploratory in nature and should be approached with an open but discerning mind, recognizing that the quest for understanding the universe is an ever-evolving journey.

Don't miss out!

Visit the website below and you can sign up to receive emails whenever SANDEEP CHAVAN publishes a new book. There's no charge and no obligation.

https://books2read.com/r/B-A-EPGPC-KYFYF

BOOKS 2 READ

Connecting independent readers to independent writers.

Did you love *The Cosmic Breath*? Then you should read *The Blueprint of the Universe* by SANDEEP CHAVAN!

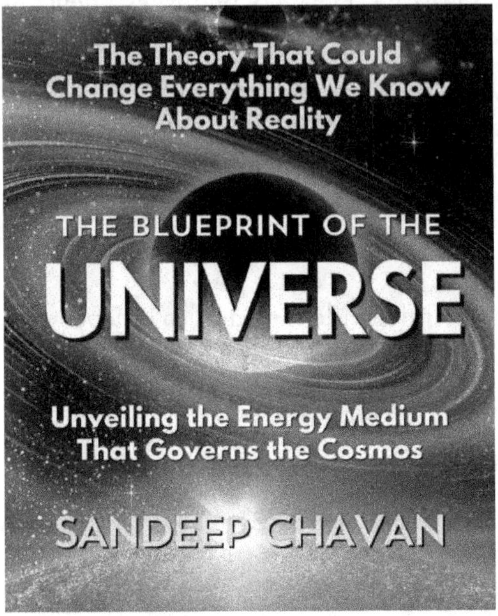

The Blueprint of the Universe: *The Theory That Could Change Everything We Know About Reality* by **Sandeep Chavan**

What if **gravity** isn't a fundamental force but an effect of **structured energy interactions**? What if **dark matter** and **dark energy** aren't mysterious substances but emergent properties of a hidden **universal energy medium**? What if the **Big Bang** wasn't the beginning but a **phase transition** in an eternal energy cycle?

In *The Blueprint of the Universe,* **Sandeep Chavan** challenges some of the biggest assumptions in physics and cosmology, introducing a **groundbreaking framework—Universal Energy Dynamics (UED)**—that redefines how we perceive **gravity, space, time, and motion**.

At the heart of this book lies a revolutionary concept: **the universe is not an empty void but a structured energy system, where every interaction follows a dynamic equilibrium**. Drawing from **Newton's mechanics, Einstein's relativity, quantum physics, and even ancient philosophical insights**, this book presents a **unified model** that connects everything—from planetary motion to black hole physics, from light propagation to the origin of consciousness.

Key Ideas Explored in the Book:

Chavan's Law of Gravity – A new perspective where **mass interacts with vacuum energy gradients**, shaping planetary motion and eliminating the need for dark matter.**The First Disturbance & Cosmic Evolution** – How **structured energy transitions** led to the formation of matter, galaxies, and even life itself.**Black Holes as Energy Regulators** – Why black holes don't "trap" energy but **redistribute it**, ensuring universal balance.**Artificial Gravity & Space Propulsion** – How **vacuum energy fields** could pave the way for the next revolution in space exploration.**The Consciousness-Energy Connection** – Why **thought, perception, and intelligence** might be emergent properties of structured energy dynamics.

This book is not just another physics theory—it's a **blueprint for the future of science**. It bridges the gaps between **classical physics, relativity, quantum mechanics, and cosmology**, offering **testable predictions** and a **new way of thinking about the cosmos.**

Who Should Read This?

Scientists, researchers, and students of physics & cosmology.Thinkers, philosophers, and those curious about the nature of reality.Space enthusiasts and AI pioneers seeking the next big breakthrough.

Are you ready to rethink everything you know about the universe? This book might just be the **missing piece** in our quest to understand existence itself.

Read more at https://www.sandeepjchavan.com/.

Also by SANDEEP CHAVAN

Echoes of War: The Russia-Ukraine Conflict and Its Global Ripples

Shattered Lives

The War Economy

Borders Redrawn

Arms and Allies

Collateral Damage

Justice in the Ashes

Somewhere in the Darkness of the Cosmos

The Blueprint of the Universe

The Cosmic Breath

The Eternal Sage Chronicles

The Sage and the Sovereign

The Family Pillar

The People's Flame

The Warrior's Soul

The Fundamental Theory of Life & Existence

Energy in Action: The Soul Demystified

The Active Mind

Energy in Connection

Death: The Disconnection

Ultra Modern Life Philosophy

The Duality of Laughter and Tears

Compare to Accept

Standalone

It's Not AI, It's AHI - Amplified Human Intelligence

The Decision Paradox

The Four Sapiens

Malicious Script of Indian Polity

The IIT Legacy & Global Impact

The IIT Legacy & Global Impact

You are just a Version of Your Original

Win the Game You Didn't Choose

Always Keep Your Bags Packed

Reveal to Shield - Learn the Game of Facades

Engage Beyond Elections

Culture, Identity & Change: The Evolution of Indian Society

The Triangular Dynamics

Never Trust Without Doubt

If You Can't Change the Country, Leave the Country

Live Fearless with Fear

India: The Investment Magnet

Realignment
Powering a Billion Dreams
Universal Skill
Extension: From Brush to Algorithm
Human Side of Einstein
The Future Knows No Lines
Let's Bring the Generations to the Table
Unified Energy Dynamics
A Beat That Never Dies
Strategic Horizons
Right From Scratch
The Forgotten Innovators

Watch for more at https://www.sandeepjchavan.com/.